Springer Water

The book series Springer Water comprises a broad portfolio of multi- and interdisciplinary scientific books, aiming at researchers, students, and everyone interested in water-related science. The series includes peer-reviewed monographs, edited volumes, textbooks, and conference proceedings. Its volumes combine all kinds of water-related research areas, such as: the movement, distribution and quality of freshwater; water resources; the quality and pollution of water and its influence on health; the water industry including drinking water, wastewater, and desalination services and technologies; water history; as well as water management and the governmental, political, developmental, and ethical aspects of water.

Muhammed Atamanalp · Gonca Alak · Arzu Uçar ·
Veysel Parlak
Editors

Aquatic Toxicology in Freshwater

The Multiple Biomarker Approach

Springer

Editors
Muhammed Atamanalp
Department of Aquaculture
Faculty of Fisheries
Atatürk University
Erzurum, Türkiye

Gonca Alak
Department of Seafood Processing Technology
Faculty of Fisheries
Atatürk University
Erzurum, Türkiye

Arzu Uçar
Department of Aquaculture
Faculty of Fisheries
Atatürk University
Erzurum, Türkiye

Veysel Parlak
Department of Basic Sciences
Faculty of Fisheries
Atatürk University
Erzurum, Türkiye

ISSN 2364-6934 ISSN 2364-8198 (electronic)
Springer Water
ISBN 978-3-031-56668-4 ISBN 978-3-031-56669-1 (eBook)
https://doi.org/10.1007/978-3-031-56669-1

This Springer imprint is published by the registered company Springer Nature Switzerland AG
The registered company address is: Gewerbestrasse 11, 6330 Cham, Switzerland

Contents

List of Figures

List of Tables

Chapter 1
History of Aquatic Toxicology

Arzu Uçar

Abstract The history of toxicology provides a fascinating insight not only into the development of the science of toxicology, but also into its changing approach to disease prevention for the protection of public health. It would not be an exaggeration to say that the knowledge about the toxic/healing properties of minerals, plants, and animals that have shaped civilization for thousands of years, toxic substances and their uses is as old as human history. Throughout the ages, the science of toxicology has provided information that has shaped and guided society, laying the foundations for modern toxicology. The science of toxicology has been built sometimes on the findings obtained as a result of basic experiments, sometimes by chance discoveries. This process, which is similar to other branches of science, has required more emphasis as it affects the health of living things in dimensions leading up to death. This chapter examines the development of the discipline of toxicology and its impact on humanity, highlighting key milestones and discoveries related to toxicology. In addition to shedding light on the beginnings of toxicology, it provides a brief overview of the story of aquatic toxicology up to the present day, taking into account the availability of xenobiotics to aquatic life and the characteristics of the aquatic environment that determine the fate of foreign chemicals within the organism.

Keywords Aquatic toxicology · History · Poison · Freshwater toxicology · Dose

List of Abbreviations

BC	Before christ
FDA	Food and drug administration
IPCS	International program on chemical safety
WHO	World health organization

A. Uçar (✉)
Department of Aquaculture, Faculty of Fisheries, Atatürk University, Erzurum, Türkiye
e-mail: arzuucar@atauni.edu.tr

M. Atamanalp et al., *Aquatic Toxicology in Freshwater*, Springer Water,
https://doi.org/10.1007/978-3-031-56669-1_1

PCB	Polychlorinated biphenyls
DDT	Dichloro diphenyl trichloroethane
US EPA	US Environmental Protection Agency
EEC	European economic community
DDT	Diklorodifeniltrikloroetan
APHA	American public health association
ASTM	American society for testing and materials
MAFF	Ministry of agriculture fisheries and food
PARCOM	Paris commission
SETAC	Society of environmental toxicology and chemistry
WRC	Water research centre

1.1 Introduction

Toxicology is defined as the qualitative/quantitative study of the positive/negative effects of physical, chemical and biological agents observed as structural and functional changes in living organisms. While effects can be observed either positively or negatively, this discipline focuses more on negative effects. In the light of the data obtained from research in this field, the protection of all living things, especially humans, from harmful effects is the main purpose and duty of this branch of science. In this branch of study, the determination of the safety of chemical substances developed in the highly improved chemical industry and new ones added every day is also accepted within the job description (Wexler and Hayes 2019). It is a dynamic, developing and predictive science that never loses its relevance despite its long history.

Toxicology has a multidisciplinary backbone with dozens of disciplines such as biology, chemistry, physics, anatomy, physiology, pathology, psychology, zoology, pharmacology, genetics, biochemistry, statistics and mathematics. The purpose of the researcher determines which one(s) of these disciplines will act in common with the properties of the chemical or living organism used and the test methods utilized. When we want to simplify the definition, toxicology can be defined as the study of the interaction of living organisms with substances, toxins, xenobiotics or stressors (Nepovimova and Kuca 2019). A more descriptive definition of toxicology is the study of the nature and mechanisms underlying the direct or indirect toxic effects of substances such as biological, chemical, physical, genetic or physiological agents on living organisms and other biological systems. More specifically, toxicology is concerned with the chemical and physical properties of poisons, their physiological or behavioral effects on living organisms, qualitative/quantitative methods for their analysis in biological and non-biological materials, and the development of procedures for the treatment of poisoning (Wexler and Hayes 2019). Toxicology also deals with the quantitative or qualitative assessment of adverse effects related to the concentration, dosage, duration and frequency of repetition to which organisms are exposed (Lee 2017).

1.1.1 Story of Aquatic Toxicology

The history of toxicology is likely to be as old as human history: The first man had to learn to "distinguish between what is good for food and what is not", to learn about the healing or harmful effects of plants and other materials as he explored his environment and searched for food. He observed that the bites of certain insects/reptiles caused illness or death and found the necessity to avoid them in the first place. The doctrine of not consuming the plant that harmed him or the efforts to destroy the creatures that have negative effects on health can be considered as the foundations of this science. This leads to the idea that knowledge of animal poisons and plant extracts probably predates recorded history (Wexler 2014).

1.1.2 Middle Ages

When the recorded history of toxicology is investigated, it is seen that 'hemlock', a plant-based poison, was used as an official state poison in ancient Greeks. In addition, it is found in written sources that Socrates (470–399 BC) was sentenced to death with "hemlock". In 132–163 BC, experiments on prisoners in the Pontic state of Pontus led to the discovery of an antidote to a wide range of toxic substances and experimentally demonstrated that humans became immune to it.

In the continuation of the written records of the science of toxicology, it has been observed that the first law for poisons was drafted in Ancient Rome, called "Lex Cornelia" (82 BC) (Sánchez 2014). It is seen that lead poisoning was observed by the lead-lined cistern and pipes contaminating drinking water by the architect Vitrivius Pollio (80–70 BC), and the potential health hazards of mining were reported for the first time in history.

Ibn Sina's "Kitab al Kanun Fit Tibb" (known as Avicenna in the West), an important step in the history of toxicology, mentions important antidotes against different poisonings (3). The Ebers Papyrus (1500 BC), one of the oldest known records, contains information on many known poisons, including plants such as hemlock, aconite, opium and metals such as lead, copper and antimony. The Book of Job (ca. 1400 BC) mentions poisoned arrows (Job 6:4), while Hippocrates (400 BC) introduced many poisons, adding a number of principles of poison and clinical toxicology related to therapy and bioavailability in overdose. In addition, he reported the occurrence of poisoning in workers working in lead mines and laid the foundation of industrial toxicology by revealing the relationship between lead poisoning and occupational diseases.

Poisons identical to today's arsenic were made, such as Cantarella, allegedly used during the reign of Pope Alexander VI and made by the Borgias, a medieval family, and these poisonings are reported to have been very important in the social and political life of Italy. Catherine de Medici was another important figure who practiced poisoning methods in Italy (Preziosi et al. 2003).

Mattieu Joseph Bonaventura Orfila (1787–1853), recognized as the founder of modern toxicology, was the first to explain the relationship between the chemical and biological properties of poisons. In his work "Traite de toxicologic", there is information revealing that poisons are collected in many organs after being absorbed from the digestive system. Orfila, who developed many methods for the identification of toxic substances, also emphasized the relationship between chemistry and forensic medicine, two different disciplines, in his work "Legons de medecine Legale". This research was the first time that "analytical toxicology" and "forensic toxicology" were clearly defined. Orfila's research and books have been taught in the field of forensic medicine all over the world and his "A Treatise on Poisons" has been published many times in Europe and America (Vural 2005).

Theophrastus (370–286 BC), a student of Aristotle, made numerous references to poisonous plants in De Historia Plantarum. Dioscorides, a Greek physician at the court of the Roman emperor Nero, made the first attempt to classify poisons into plant, animal and mineral poisons in his De Materia Medica, which refers to about 600 plants.

Although the history of poisons dates back to the earliest times, the science of toxicology has important traces of Paracelsus (1493–1541) and Orfila (1757–1853). The sixteenth-century physician Paracelsus (1493–1541), who was clearly aware of the dose–response relationship, wrote: "All substances are poison, the difference between medicine and poison is the dose. He has a determination that states, "The right dose separates the poison from the drug". With this determination, he revealed that the "therapeutic" and "toxic" properties of the chemical substance should be separated. Paracelsus claimed that specific chemicals are actually responsible for the toxicity of a plant or animal poison, that there is no non-toxic substance in the world where the therapeutic and toxic properties of substances cannot be distinguished down to a single parameter, namely the dose (Nepovimova and Kuca 2019). This assumption is still one of the pillars of modern toxicology. It also revealed that the body's response to these chemicals depends on the dose taken, that low doses of a substance can be harmless or even beneficial, while larger doses can be toxic, which points to the dose–response relationship (hormesis) known in toxicology (Öztürk et al. 2022).

The foundations of the development of toxicology as a distinct branch of science were laid in the eighteenth and nineteenth centuries. The detection of poisons started with the investigation of the toxic effects of drugs and chemicals in animals. The mystical approach to poisons was replaced by a scientific and realistic approach and was first systematized by Matthieu Orfila (1787–1853) in the nineteenth century. Accordingly, the functions of the toxicologist have traditionally been to identify poisons, search for antidotes and other ways of treating toxic injuries (Britannica 2018). In the late eighteenth century, Matthieu Joseph Bonaventure Orfila, a Spanish physician working at the University of Paris, first produced a systematic correlation between the chemical and biological properties of the poisons of the time. Using autopsy materials for poisons and associated tissue damage, he demonstrated the effects of toxic substances and compounds on specific organs. He stated that research of this scale is too important to be left to people who do not have sufficient knowledge and experience, and emphasized that toxicology should be a separate science (Langman and Kapur 2006).

Research on the analysis of arsenic, the most widely used poison of the 1700s, formed the basis of analytical toxicology. During this period, research into the mechanisms of toxic action was concentrated in France and Germany. Froncois Magendie (1783–1855) investigated the mechanisms of action of emetine, strychnine and cyanide, and his student Claud Bernard (1813–1878) conducted important research to understand the mechanisms of action of carbon monoxide and curare.

Louis Lewin (1850–1929) was the first scientist to emphasize the difference between the pharmacological and toxicological mechanisms of drugs.

By the 1800s, Louis Lewin (1850–1929) was the first scientist to emphasize the difference between the pharmacological and toxicological mechanisms of drugs. After these years, toxicology was acknowledged as a science and toxic effects mechanisms such as acute and chronic toxicity, neurotoxicity, selective toxicity, internal effects of radioactivity were started to be studied by Carl Voegtlin et al. (1923) and Rudolf Peters et al. (1945). Biological effect (dose–response) relationship has also started to be evaluated by examining drugs and chemical substances in blood and urine. With the introduction of instrumental analysis into toxicology, analysis of substances and compounds in tissues and other biological materials began in the 1940s. In this period, developments in the field of toxicology enabled toxicology to work together with other disciplines and continued to develop until today.

More than 2000 years ago, Aristotle studied the effect of salinity by placing freshwater animals in seawater. This experimental practice was later considered as a toxicity study. With this study, the effect of a substance on an organism aroused the curiosity of many physiologists. However, mankind first used chemicals and then studied their effects. Although the harmful effects of these substances on many organisms are known, toxicity studies in the disciplinary sense were not recorded much until the 1940s and 1950s. Before World War II, some researchers studied the toxic effects of metals, especially in fish. After the war, many toxicity laboratories were established in England, the United States and Canada.

Investigations of toxic effects at the molecular level and the explanation of the mechanisms of chemical carcinogenesis are seen in the 1940s. Miller identified mixed function oxidases in the endoplasmic reticulum and started research on cytochrome p450 oxidase enzymes. In 1947, Williams published "Mechanisms of Detoxication" in which he described the many mechanisms involved in the detoxification of toxins. In the United States, the Food and Drug Administration (FDA) legalized a toxicology and safety assessment program in 1955 to ensure the safety of food, drugs and cosmetics. The 1960 "thalidomide disaster" brought to the forefront the importance of toxicological research in pharmaceuticals.

In parallel, WHO initiated the publication of an International Program on Chemical Safety (IPCS) and a series of Environmental Health Criteria, and organochlorine insecticides were recognized in the 1960s as contaminants of water systems worldwide in the chromatograms of environmental analyses (Vasseur et al. 2021). With this regulation, the classification of pollutants has been formalized and a serious record has been kept on the issues to be complied with.

These negativities have shown that the studies to be carried out by this branch of science will be prepared worldwide. These events, which are defined as negative, have been positive for the toxicology discipline and have revealed areas of study. The effects of mercury, cadmium and PCB poisoning on humans in Japan, the observation of toxic effects due to DDT (dichloro diphenyl trichloroethane) in birds and fish in the Great Lakes region of North America (Peakall 1994), ecotoxicology, which developed with the pollution of the environment by industrial and agricultural pollutants (Ramade 1992), revealed the first evidence in the 1960s (Narbonne 2021). This area of work has contributed to raising awareness of the environmental trajectories and impacts of pollutants and the negative side effects of the chemicals used.

The environmental science book "Silent Spring" written by Rachel Carson in 1962 focused on the effects of pesticides, ecological problems, pollution of surface and ground water, inadequate water treatment facilities, persistence and transfer of chlorinated hydrocarbons, the use of natural products for biological control, human safety and the harm of pesticide resistance on target organisms. Focusing on water pollution, this book's explanations of pesticide impacts on a broad scale have brought what was known to environmentalists to the forefront of the political, commercial, scientific and public sectors of society and, despite controversy, have had a long-lasting impact (Rattner 2009).

The first decade of the 1970s spurred a strong international mobilization for pollution prevention, regulation and research. The US EPA (US Environmental Protection Agency) was established in 1970, Environment Canada in 1971, leading to the banning of most organochlorine insecticides by the EPA in 1972. By Decision of December 21, 1978, the European Economic Community (EEC) issued EEC Directive 79/117/CEE banning (as of 01/01/1981) phytosanitary products containing aldrin, dieldrin, chlordane, DDT, endrin, heptachlor, hexachlorobenzene, alkyl and aryl. At the same time, EEC Directive 79/831/CEE amending the 6th Directive 67/548/CEE on the classification and labeling of chemicals addressed the regulation requiring a notification dossier for new chemical substances since September 18, 1981.

After the 1970s, environmental pollutants came to the forefront of toxicology and a war on toxic substances was initiated by an environmental law in the United States. In the late 1970s and early 1980s, toxicity studies increased tremendously, and the American Public Health Association (APHA), the American Society for Testing and Materials (ASTM), the U.S. Environmental Protection Agency (U.S. EPA), the U.S. The U.S. Army Corps of Engineers of Materials the UK Ministry of Agriculture Fisheries and Food (MAFF), the Paris Commission (PARCOM), the Society of Environmental Toxicology and Chemistry (SETAC) and the Water Research Centre (WRC) have developed a range of standard methods.

1.1.3 The Story of Toxicology's Subdivision

Chemical ecology and ecotoxicology is the field of research that emerged between the 1950s and 1970s following increased awareness of information or chemicals that affect biotic interactions or the ecological and ecosystem consequences of anthropogenic pollutants.

How hazardous chemical, biological and physical agents occurring in soil, air and water environments can affect individual living organisms is the central question of Environmental Toxicology, a multidisciplinary scientific field. Similarly, the effects of such agents on individuals, populations, communities and ecosystems form the basis of the multidisciplinary field of Ecotoxicology, which integrates ecology and toxicology and is a sub-discipline of Environmental Toxicology. Therefore, questions about the meaning of the word "clean" are a major challenge for Environmental Toxicology and Ecotoxicology. Toxicological predictions of both research fields are of great importance for national economies and ecological sustainability (Agathokleous and Calabrese 2020).

Metals, pesticides and personal care products are among the most emphasized pollutants that can interfere with the chemical ecology of organisms. Given the widespread environmental pollution of the Anthropocene epoch, it seems important to consider disturbances such as "information pollution" in environmental risk assessment. A wide range of anthropogenic or natural pollutants are involved in biotic interactions as they mediate this degradation. Chemical ecology can provide new response factors that can help determine the non-lethal effects of pollutants. Furthermore, discovering natural, non-toxic alternatives to currently used biocides can help with risk management. The link between both disciplines should be strengthened, as both are already multidisciplinary fields and developing common themes between chemical ecology and ecotoxicology can provide a deeper understanding of ecological processes. It could also help to achieve the objective of the 2020 European chemicals strategy for sustainability towards a non-toxic environment (Gross 2022).

1.1.4 Modern Toxicology

Modern toxicology is characterized by sophisticated scientific research and assessment of toxic exposures. As the chemical industry develops new chemicals, this branch of science will continue to develop in a wide range of fields. Modern toxicology has led to the discovery of DNA and various biochemicals that maintain cellular functions, and has sought to expand knowledge of toxic effects on organs and cells. From this point of view, it is recognized that almost all toxic effects are caused by changes in specific cellular molecules and biochemicals. Toxic agents are classified in various ways and there is no single classification that applies to the entire spectrum of toxic substances. Therefore, purpose-built classification systems and their combinations may provide the best system (Gupta and Gupta 2020).

Aquatic ecotoxicology has become more complex than the simple application of laboratory analysis to assess the effects of a specific compound on a selected bioindicator, an organism or group of organisms, and its biochemical, physiological or histological indicators or biomarkers (Forbes et al. 2006; Markert et al. 2003). While this one-to-one approach is still valid, especially during research prior to the release of new xenobiotics into the industrial chain, toxicity experiments have become useful in the assessment of mixtures of multiple substances in complex aquatic environments. Recent ecotoxicology research focuses on numerous biomarkers and organisms at different levels of the trophic chain, populations, communities and ecosystem (Uçar et al. 2020, 2021, 2022, 2023; Alak et al. 2017, 2018, 2019, 2020, 2021, 2022; Atamanalp et al. 2021, 2022, 2023; Parlak et al. 2022). Multidisciplinary studies have contributed a lot to the field of toxicology with the developments in each discipline. Today, new ecotoxicity experiments are being designed that take into account more variables known to affect organismal responses. It will continue to develop with the increasing number of pollutant sources every day. Studies involving discharge regime or dosage mode (continuous, intermittent, episodic, spraying and accidental spills) aiming to simulate different events such as sewage and industrial wastewater discharges (point sources) are being conducted (de Sa Salomao et al. 2020).

One of the sub-disciplines of toxicology that has become a star in recent years is "aquatic toxicology". As the name suggests, it studies on living things whose nature is water. Aquatic toxicology and environmental risk assessments of chemicals or compounds have been included as an important component in some parts of the EU chemicals legislation. Aquatic toxicology can be defined as the study of toxins and their adverse effects on aquatic ecosystems at different spatial, temporal and organizational scales. Since aquatic environments contain thousands of species, each of these species can respond to toxic substances in many ways, and the interactions between these species contain information useful for toxicology. This provides an opportunity to assess the state of aquatic systems, monitor trends over time relative to conditions, diagnose the cause of problematic systems, guide efforts to correct damage, and predict the consequences of proposed human actions so that the ecological consequences of these actions can be assessed before damage occurs.

The fact that every chemical can have a toxic effect in a dose-dependent manner makes the field of toxicology a very wide area, including pharmaceuticals, cosmetics, pesticides, food additives, household cleaners and industrial chemicals. All these chemicals are also called "xenobiotics", meaning foreign to the organism.

In recent decades, increasing inputs of anthropogenic pollutants and other stressors, such as habitat destruction, have led to drastic changes in aquatic ecosystems. Accordingly, increased attention has focused on:

(a) accumulation and toxic effects of pollutants on the survival of aquatic organisms; (b) the uptake and accumulation of pathogenic organisms and chemical contaminants in water supplies intended for human consumption; and (c) the release of biodegradable organic matter and nutrients that can cause localized eutrophication, organic enrichment and oxygen depletion under dormant conditions (Malins and Ostrander 1991).

This chapter addresses the aquatic organism-xenobiotic interaction, as well as aspects of the interaction that determine environmental trajectory and toxicity. Contamination of aquatic environments with pesticides and fertilizers from agricultural fields, as well as airborne deposition of mercury, polychlorinated biphenyls and polycyclic aromatic hydrocarbons, are examples of the introduction of toxic substances into aquatic systems, often by non-point sources. Groundwater contamination arises from many sources, including natural sources such as septic tanks, landfills, hazardous waste sites, pipelines, irrigation, pesticide applications, wells, saltwater intrusion or natural seepage. Those who first start studying aquatic toxicology encounter two important terms: "water pollution" and "water quality". Water pollution can be defined as the deterioration of the physical, chemical and biological properties of water. Water quality can be expressed as the suitability of water for the purpose for which it will be used. For example, the term water quality has a very different meaning for a fish farming water than for a car wash. Water quality can affect the availability, chemical form and ultimately the toxicity of chemicals in the water environment. Sediments are the primary repositories for abiotic and biotic materials in the aquatic environment. Water hardness is influential in the distribution and toxicity of various water pollutants. Ingestion of pollutant-associated particles can provide another important pathway of pollutant assimilation in sediment-feeding benthic organisms (Kleinow and Goodrich 2019).

Aquatic toxicology is focused on research applying toxicological, physiological and biochemical approaches in different taxa, including fish (mainly trout and halibut) and molluscs. Metals and pesticides that cause pollution in freshwaters have formed the main components of toxicology. In this process, studies on the exposure of aquatic organisms to chemicals and the responses of organisms to these exposures began, and important response pathways mediated by toxic substances were outlined without understanding the molecular details. For example, the dissolution of petroleum mixture (Frysinger et al. 2003) and the application of molecular and genomic techniques have led to significant advances in our understanding of how chemical exposure alters gene expression (Wang et al. 2010). In addition, significant efforts have been made and much progress has been made in identifying and characterizing the genes and proteins involved in the mechanisms of toxicity. Numerous studies have also addressed the comparative biology of chemical effects, transcription factors, biotransformation enzymes and transporters and how these vary between taxa (Goldstone et al. 2006; Uçar et al. 2020, 2021, 2022, 2023; Alak et al. 2017, 2018, 2019, 2020, 2021, 2022; Atamanalp et al. 2021, 2022, 2023; Parlak et al. 2022).

1.1.5 Current Aquatic Toxicology

Despite the hard work of many researchers, there are large gaps in the knowledge of toxicological effects in freshwater systems. Toxicologists are now using in vitro methods in metabolism studies to better elucidate chemical bioaccumulation in fish, an important component of environmental risk assessments for ecosystems and human health (i.e. consumption of contaminated fish). As science continues to advance, so does the ability to correlate chemical concentrations that induce responses in vitro with exposure levels in vivo and adverse human and ecological outcomes. Because of their molecular, biochemical and physiological similarities to humans, mammals (especially rodents) have historically been used in biomedical and, more specifically, toxicology research. More recently, however, various fish species have been increasingly used as vertebrate alternatives to rodents in toxicology, pharmacology and etiology studies of human diseases. The species used in aquatic toxicity tests will be discussed in detail in the relevant section. Although the number of aquatic species utilized is large, zebrafish (*Danio rerio*) has come to the forefront due to its unique characteristics. In particular, zebrafish are used in the suitability screening of newly developed chemicals/drugs and they can proceed to the next stages for the suitability of these products to be used by mankind. These toxicity tests in fish are as important as the support of clinical tests in the next steps. In other words, aquatic toxicology is of serious importance for mankind to benefit from the new products developed.

Aquatic organisms, especially fish, are now widely used in laboratory research supported by well-developed comparative biology resources. Similar to in vitro tools, fish are used as model organisms in that they are cost and time effective relative to mammals, help mitigate animal welfare concerns, fish develop/reproduce faster than mammals, and provide clear benefits during high-throughput screening of numerous chemicals.

The ecological importance of fish among aquatic organisms and their use as model organisms, as well as the evolutionary relationships between fish and mammals, allow fish models to provide advantages in molecular mechanistic studies. As mentioned above, several recently developed technical resources (such as sequenced genomes, DNA libraries and available antibodies) further support the use of fish for biomedical research. The fully-established genome databases of zebrafish (*Danio rerio*) and fathead minnow (*Pimephales promelas*) indicate high evolutionary conservation compared to the human genome. For example, 70% of human protein-coding genes are related to genes found in zebrafish, and many mutant phenotypes have been reported to resemble clinical human diseases (Brooks et al. 2020).

The field of aquatic toxicology utilizes a combination of bioinformatics and biotechnological approaches that provide important information on the pathway to orchestrate a new design (Hartung and Leist 2008). Three important technologies developed in the last decade, such as 'omics' (such as genomic and proteomic analyses), imaging techniques and robotized testing platforms, have entered the field of toxicology (Pognan 2004). Omics technologies and imaging methods enable the collection of massive sets of information about a single compound, while assay platforms enable the testing of large numbers of substances by analyzing large numbers of samples under standardized conditions. Advances in bioinformatics and in silico modeling can also be leveraged to integrate information from other areas of the life sciences (Lotti and Nicotera 2002). Such integration of knowledge will be particularly important for the investigation of cellular pathways and should allow cross-applications of ideas between toxicology and basic sciences. The combination of biochemical knowledge of cell pathways with genomics, proteomics and metabonomics (the study of metabolic responses to environmental factors, drugs and diseases) is already advancing as systems biology, and systems toxicology is a new subfield of this field (Hartung 2009).

1.2 Conclusion

The goals and challenges faced today reveal the need to build on the achievements made in the field of aquatic toxicology in freshwaters, as well as to address the pressing issues brought about by technology and globalization. It is important for the development of aquatic toxicology to examine what challenges await freshwater ecosystems in the coming years, and to explore multiple, holistic approaches and methods to tackle the ever-increasing sensitivity and complexity of chemicals and their effects.

The scientific adventure that started with the main title of toxicology has branched off into secondary roads and nowadays has reached the paths of freshwater toxicology.

Acknowledgements Many thanks to my colleagues from Fisheries Faculty who supported me in the research for their help Atatürk University for their support in my academic studies.

References

Agathokleous E, Calabrese EJ (2020) Environmental toxicology and ecotoxicology: how clean is clean? Rethinking dose-response analysis. Sci Total Environ 746:138769. https://doi.org/10.1016/j.scitotenv.2020.138769

Alak G, Yeltekin AÇ, Tas IH, Ucar A, Parlak V, Topal A, Atamanalp M (2017) Investigation of 8-OHdG, CYP1A, HSP70 and transcriptional analyses of antioxidant defence system in liver tissues of rainbow trout exposed to eprinomectin. Fish Shellfish Immunol 65:136–144. https://doi.org/10.1016/j.fsi.2017.04.004

Alak G, Ucar A, Yeltekin AÇ, Çomaklı S, Parlak V, Taş IH, Türkez H (2018) Neuroprotective effects of dietary borax in the brain tissue of rainbow trout (*Oncorhynchus mykiss*) exposed to copper-induced toxicity. Fish Physiol Biochem 44:1409–1420. https://doi.org/10.1007/s10695-018-0530-0

Alak G, Yeltekin AÇ, Uçar A, Parlak V, Türkez H, Atamanalp M (2019) Borax alleviates copper-induced renal injury via inhibiting the DNA damage and apoptosis in rainbow trout. Biol Trace Elem Res 191:495–501. https://doi.org/10.1007/s12011-018-1622-5

Alak G, Parlak V, Ucar A, Yeltekin AC, Ozgeris FB, Caglar O, Turkez H (2020) Oxidative and DNA damage potential of colemanite on zebrafish: Brain, liver and blood. Turk J Fish Aquat Sci 20(8):593–602. https://doi.org/10.4194/1303-2712-v20_8_02

Alak G, Ucar A, Parlak V, Yeltekin AÇ, Özgeriş FB, Atamanalp M, Türkez H (2021) Antioxidant potential of ulexite in zebrafish brain: assessment of oxidative DNA damage, apoptosis, and response of antioxidant defense system. Biol Trace Elem Res 199:1092–1099. https://doi.org/10.1007/s12011-020-02231-7

Alak G, Uçar A, Parlak V, Atamanalp M (2022) Identification, characterisation of microplastic and their effects on aquatic organisms. Chem Ecol 38(10):967–987. https://doi.org/10.1080/02757540.2022.2126461

Atamanalp M, Kırıcı M, Köktürk M, Kırıcı M, Kocaman EM, Ucar A, Alak G (2023) Polyethylene exposure in rainbow trout; suppresses growth and may act as a promoting agent in tissue-based oxidative response, DNA damage and apoptosis. Process Saf Environ Prot 174:960–970. https://doi.org/10.1016/j.psep.2023.05.005

Atamanalp M, Köktürk M, Uçar A, Duyar HA, Özdemir S, Parlak V, Alak G (2021) Microplastics in tissues (brain, gill, muscle and gastrointestinal) of *Mullus barbatus* and *Alosa immaculata*. Arch Environ Contam Toxicol 81:460–469. https://doi.org/10.1007/s00244-021-00885-5

Atamanalp M, Türkez H, Yeltekin AÇ, Özgeriş FB, Ucar A, Çağlar Ö, Alak G (2022) Borax relieved the acrylamide-induced hematotoxic, hepatotoxic, immunotoxic and genotoxic damages in rainbow trout by regulating apoptosis and Nrf2 signaling pathway. Comp Biochem Physiol C: Toxicol Pharmacol 259:109396. https://doi.org/10.1016/j.cbpc.2022.109396

Britannica T Editors of Encyclopaedia (2018, June 25). Toxicology test. Encyclopedia Britannica. https://www.britannica.com/science/toxicology-test

Brooks BW, Sabo-Attwood T, Choi K, Kim S, Kostal J, LaLone CA, Zhang X (2020) Toxicology advances for 21st century chemical pollution. One Earth 2(4):312–316. https://doi.org/10.1016/j.oneear.2020.04.007

Forbes VE, Palmqvist A, Bach L (2006) The use and misuse of biomarkers in ecotoxicology. Environ Toxicol Chem Int J 25(1):272–280. https://doi.org/10.1897/05-257R.1

Frysinger GS, Gaines RB, Xu L, Reddy CM (2003) Resolving the unresolved complex mixture in petroleum-contaminated sediments. Environ Sci Technol 37(8):1653–1662

Goldstone JV, Hamdoun A, Cole BJ, Howard-Ashby M, Nebert DW, Scally M, ... Stegeman JJ (2006) The chemical defensome: environmental sensing and response genes in the Strongylocentrotus purpuratus genome. Dev Biol 300(1):366–384

Gross EM (2022) Aquatic chemical ecology meets ecotoxicology. Aquat Ecol 56(2):493–511. https://doi.org/10.1007/s10452-021-09938-2

Gupta PK (2020) General toxicology. Problem Solving Questions in Toxicology: a study guide for the board and other examinations, pp 11–26.https://doi.org/10.1007/978-3-030-50409-0

Hartung T (2009) Toxicology for the twenty-first century. Nature 460:208–212. https://doi.org/10.1038/460208a

Hartung T, Leist M (2008) Food for thought... on the evolution of toxicology and phasing out of animal testing. ALTEX-Alternatives to animal experimentation 25(2):91–102. https://doi.org/10.14573/altex.2008.2.91

Kleinow KM, Goodrich MS (2019) Environmental aquatic toxicology. In Basic environmental toxicology. CRC Press, pp 353–384

Langman LJ, Kapur BM (2006) Toxicology: then and now. Clin Biochem 39(5):498–510. https://doi.org/10.1016/j.clinbiochem.2006.03.004

Lee BM (2017) History of toxicology. In Lu's Basic Toxicology. CRC Press, pp 3–21

Lotti M, Nicotera P (2002) Toxicology: a risky business. Nature 416(6880):481–481. https://doi.org/10.1038/416481a

Malins DC, Ostrander GK (1991) Perspectives in aquatic toxicology. Annu Rev Pharmacol Toxicol 31(1):371–399. https://doi.org/10.1146/annurev.pa.31.040191.002103

Markert BA, Breure AM, Zechmeister HG (Eds) (2003) Bioindicators and biomonitors. Elsevier

Narbonne JF (2021) History—biological basis of the use of biomarkers in ecotoxicology. In: Use of biomarkers for environmental quality assessment. CRC Press, pp 1–7

Nepovimova E, Kuca K (2019) The history of poisoning: from ancient times until modern ERA. Arch Toxicol 93(1):11–24. https://doi.org/10.1007/s00204-018-2290-0

Öztürk F, Uçar A, Atamanalp M (2022) Akuatik Toksikolojide Adaptif Yanıt: Hormesis. Ortadoğu 7. Uluslararasi modern bilimsel çalişmalar kongresi, Bayrut, Lübnan, 3 - 04 Mart 2022, ss 1–2

Parlak V, Ozgeris B, Ucar A, Cilingir Yeltekin A, Ozgeris FB, Cağlar O, Atamanalp M (2022) Hematotoxic, oxidative and genotoxic damage in rainbow trout (*Oncorhynchus mykiss*) after exposure to 3-benzoylpyridine. Toxicol Mech Methods 32(7):501–509. https://doi.org/10.1080/15376516.2022.2049413

Peakall DB (1994) The role of biomarkers in environmental assessment (1). Introduction Ecotoxicology 3:157–160. https://doi.org/10.1007/978-94-011-2346-4_10

Peters RA, Stocken LA, Thompson RHS (1945) British anti-lewisite (BAL). Nature 156(3969):616–619

Pognan F (2004) Genomics, proteomics and metabonomics in toxicology: hopefully not 'fashionomics.' Pharmacogenomics 5(7):879–893. https://doi.org/10.1517/14622416.5.7.879

Preziosi P, Dracos A, Marcello I (21 August 2003) Information resources in toxicology, Italy. Toxicology 190(1–2):35–54. https://doi.org/10.1016/S0300-483X(03)00195-1

Ramade F (1992) Precis of ecotoxicology. https://inis.iaea.org/search/search.aspx?orig_q=RN:25000546

Rattner BA (2009) History of wildlife toxicology. Ecotoxicology 18(7):773–783. https://doi.org/10.1007/s10646-009-0354-x

Sa Salomao AL, Hauser-Davis RA, Marques M (2020) Critical knowledge gaps and relevant variables requiring consideration when performing aquatic ecotoxicity assays. Ecotoxicol Environ Saf 203:110941. https://doi.org/10.1016/j.ecoenv.2020.110941

Sánchez J (2014) Classrooms, salons, academies, and courts: Mateu Orfila (1787–1853) and Nineteenth-Century French Toxicology, pp 162–186

Ucar A, Ozgeris FB, Parlak V, Yeltekin AC, Turkez H, Alak G, Atamanalp M (2022) Ulexite modulates the neurotoxicological outcomes of acetylferrocene-exposed rainbow trout. Environ Mol Mutagen 63(6):286–295. https://doi.org/10.1002/em.22498

Uçar A, Parlak V, Çilingir Yeltekin A, Özgeriş FB, Çağlar Ö, Türkez H, Alak G, Atamanalp M (2021). Assesment of hematotoxic, oxidative and genotoxic damage potentials of fipronil in rainbow trout *Oncorhynchus mykiss*, Walbaum. Toxicology Mechanisms and Methods, 31(1):73–80.https://doi.org/10.1080/15376516.2020.1831122

Uçar A, Parlak V, Özgeriş FB, Yeltekin AÇ, Alak G, Atamanalp M (2020) Determination of Fipronil toxicity by different biomarkers in gill and liver tissue of rainbow trout (*Oncorhynchus mykiss*). In Vitro Cellular Dev Biol-Animal 56:543–549. https://doi.org/10.1007/s11626-020-00480-3

Ucar A, Yeltekin AÇ, Köktürk M, Calimli MH, Nas MS, Parlak V, Alak G, Atamanalp M (2023). Has PdCu@ GO effect on oxidant/antioxidant balance? Using zebrafish embryos and larvae as a model. Chem-Biol Interact 378:110484. https://doi.org/10.1016/j.cbi.2023.110484

Vasseur P, Masfaraud JF, Blaise C (2021) Ecotoxicology, revisiting its pioneers. Environ Sci Pollut Res 28:3852–3857. https://doi.org/10.1007/s11356-020-11236-7

Voegtlin C, Smith MI, Dyer H, Thompson JW (1923) Penetration of arsenic into the cerebrospinal fluid, with particular reference to the treatment of protozoal infections of the central nervous system. Public Health Reports (1896–1970), pp 1003–1021

Vural N (2005) Toksikoloji, 2. Baskı. Ankara Üniversitesi Eczacılık Fakültesi Yayınları, s. 2–9

Wang F, Wang J, Liu X (2010) Direct evidence of a surface quenching effect on size-dependent luminescence of upconversion nanoparticles. Angew Chem Int Ed 49(41):7456–7460

Wexler P (2014) History of toxicology and environmental health. Academic Press

Wexler P, Hayes AN (2019). The evolving journey of toxicology: a historical glimpse. Casarett and Doull's toxicology: the basic science of poisons. 9th ed. McGraw-Hill Education, New York, pp 3–23

Chapter 2
Introduction to Aquatic Toxicology

Muhammed Atamanalp

Abstract In recent years, freshwater ecosystems are facing serious threats due to anthropogenic problems such as industrial wastes, agricultural activities, inadequate urban waste management and population growth. However, changes in abiotic factors such as climate change, precipitation and temperature levels directly or indirectly affect the habitats of aquatic flora and fauna, including reproduction and nutrition. It is therefore crucial to identify pollution activities, their sources, spatial distribution and fate in the aquatic environment to protect species in freshwater systems and ensure the normal functioning of ecosystems globally. These problems are among the missions of aquatic toxicology and continue to evolve with the integration of other sciences. In this chapter, the ultimate goal of toxicological research is to examine how contamination affects an ecosystem and the pathways by which toxic substances reach aquatic organisms.

Keywords Freshwater ecosystems · Pollution · Aquatic organisms · Ecotoxicological studies · Fish physiology

List of Abbreviations

BCF Bioconcentration factor
BAF Bioaccumulation factor
BMF Biomagnification factor
FAO Food and agriculture organization

M. Atamanalp (✉)
Department of Aquaculture, Faculty of Fisheries, Atatürk University, Erzurum, Türkiye
e-mail: mataman@atauni.edu.tr

M. Atamanalp et al., *Aquatic Toxicology in Freshwater*, Springer Water,
https://doi.org/10.1007/978-3-031-56669-1_2

2.1 Introduction

Pollution is environmental changes that cause direct harmful effects on all living resources, especially humans, or indirect effects on living things by affecting other components of the environment (Don-Pedro 1990). The discharge of various pollutants into aquatic environments is the result of numerous anthropogenic activities that threaten the health of living organisms and degrade the quality of water bodies by making them unfavorable (Ekubo and Abowel 2011). Anthropogenic pollution and industrial wastes (heavy metals, detergents, plastics) contribute to water pollution problems and aquatic systems deal with pollutants and their toxic effects (Hampel et al. 2015).

"Aquatic ecosystems", where biotic components interact with the abiotic components of the ecosystem, are investigated in two parts as "marine and freshwater ecosystems". Saltwater ecosystems are the largest aquatic ecosystems, covering more than 70% of the earth's surface as oceans, seas, estuaries, and coastal ecosystems. Freshwater ecosystems are lotic, lentic and wetland ecosystems covering less than 1% of the earth's surface (Bashir et al. 2020). At this stage, an important situation arises that complicates the work of aquatic toxicologists. Because each of these systems has very different components and very different properties from the others.

Seas and oceans face threats that are unique and much larger in scale than freshwater. Lakes and rivers, on the other hand, are subject to threats that are smaller in volume but more numerous in number. In terms of defense power, inland waters are weaker. They are vulnerable ecosystems due to fluctuations in flow rates, extensive urban waste, pollution pressures from industrial/agricultural activities (Ibor et al. 2019).

Freshwater ecosystems appear to be particularly at risk of global environmental change, with greater declines in biodiversity than their terrestrial or marine counterparts. This susceptibility is due to a combination of three factors: (i) freshwaters are disproportionately exposed to numerous, severe environmental and anthropogenic stressors resulting from the dependence of human populations on freshwater resources (Strayer and Dudgeon, 2010), (ii) freshwater ecosystems are biodiversity hotspots—reflecting great species richness per unit area or volume of water (Strayer and Dudgeon 2010), (iii) freshwater organisms appear to be particularly sensitive to environmental change due to the unique environmental conditions found in these systems, for example the limited availability of oxygen in the underwater environment (Forster et al. 2012). With projected increases in anthropogenic pressures on freshwater resources, the biodiversity and functioning of freshwater systems are at significant risk (Davis et al. 2015).

2.1.1 What Is Aquatic Toxicology?

Aquatic toxicology, which is included in ecotoxicology, examines the effects of industrial, human, animal, agricultural, etc. pollution elements on aquatic organisms, populations and ecosystems. This branch of science, which is too young to be compared with other branches of science, has gained a place in the scientific community with an extraordinary speed. Aquatic toxicology, defined as the study of the effects of anthropogenic materials and activities on aquatic organisms, studies at different levels from cell, individual (organism), community and ecosystems (Rand 1995). These levels also include different dimensions within themselves. For example, studies at the individual level can be conducted at different life stages such as eggs, larvae and adults. The fact that there are so many research topics and stages shows the complexity of this branch of science. All of these have to be studied because toxic substances accumulated in the aquatic environment, fish and other aquatic organisms pose a danger to both these aquatic organisms and consumers.

Ecotoxicological studies have focused on assessing and predicting how pollutants can affect organisms and ecosystems holistically, assessing and predicting their harmful effects. The overall goal is to identify early signals of pollution, define environmentally safe concentrations of compounds and stop environmental degradation (Hellou 2011). Practically, it requires a series of standardized short-term (i.e. "acute") exposure tests on different organisms, the results of which are combined to assess the stressful, harmful or lethal effects of a given compound (Klaminder et al. 2014). Detailed information on these tests is provided in the "Toxicity tests" section of the book. For fish-based tests in ecotoxicology, research on the effects of (1) acute and full life-cycle and multi-generational exposures, (2) specific biological effects versus traditional endpoints, and (3) a growing understanding of single and multiple stressor effects is critical (Thoré et al. 2021). These tests were developed to facilitate response comparison across species and substances (Rudén et al. 2017). In this way, a large amount of data has been generated on organismal responses to pollutant exposure (Brady et al. 2017).

Aquatic toxicology is focused on how contaminants reach organisms, how they cause adverse effects, as well as the tools and methods used to predict toxicity in aquatic vertebrates/invertebrates and primary producers (plants, algae) (Samantha et al. 2020). Throughout the history of the world, aquatic ecosystems have formed both a mosaic and a diversity of habitats, from freshwater springs, rivers, lakes, wetlands of continents/islands to estuaries, shallow coastal habitats, reefs and seas. The importance of fish in the human diet is recognized and is increasing day by day with the awareness of healthy nutrition. This living group, which provides various ecosystem services to humans, is one of the most important natural resources of the world as they contribute to the health, well-being, cultural identity and economies of societies (Hughes 2015). With more than 33,500 species already identified, fish make up a significant proportion of the biomass of aquatic ecosystems, with more fish species in freshwater than in marine ecosystems. However, population declines, mainly caused by habitat loss and degradation, invasive species, overfishing, pollution and climate change, increase the risk of global or local extinction of species. In the current century, marine and freshwater organisms are reported to face a series of environmental challenges unprecedented in human history (Fricke et al. 2018). Freshwater ecosystems, which harbor high levels of biodiversity, are being altered through a variety of anthropogenic activities (Reid et al. 2018) and are threatened by a range of stressors that, individually or in combination, can exert pressure on biological communities (Reid et al. 2018; Namba et al. 2020).

2.1.2 Classification of Pollutions in Aquatic Systems

There is no single classification applicable to the entire spectrum of toxic substances; therefore, combinations of classification systems based on several factors can provide the best rating system. Classification based on sources of toxic substances (plant toxins, animal toxic substances, mineral toxic substances, synthetic toxic substances), physical state of toxic substances (liquid, solid, gas, dust toxic substances), target organ/system (neurotoxic, hepatotoxic, nephrotoxic, pulmotoxic, hematotoxic, dermatotoxic substances), chemical structure (metals, nonmetals, acids and alkalis), organic toxic substances (carbon oxides, carbonates and metallic carbides and carbon compounds other than cyanides), analytical behavior of toxic substances (volatile toxic substances, extractive toxic substances, metals and metalloids), type of toxicity (acute, subacute, chronic, etc.), toxic effects (carcinogens, mutagens, teratogens, clastogens), area of use (insecticides, fungicides, herbicides, rodenticides, food additives, etc.), mechanism of action, environmental and public health is taken into account in the field of toxicology (Gupta 2016).

In order to improve the ability to relate chemical exposure to toxicological and ecological effects, aquatic toxicology is having to move from observing what chemical concentrations cause adverse effects to explanatory approaches about biological processes and mechanistic pathways leading to exposure. It is imperative that this very broad field of study explores genetic, physiological and ecological characteristics of the biota, as well as (i) accounting for species differences in toxic responses on the basis of evolutionarily evolved diversity of phenotypic vulnerability to environmental stressors, (ii) biological reading of diverse biological response profiles to prioritize chemicals, categorize them according to their mode of action and guide targeted toxicity assessment; (iii) prediction of ecological consequences of toxic exposure from knowledge of how biological processes and phenotypic traits lead to effect propagation across levels of the biological hierarchy; and (iv) exploration of concepts for assessing the cumulative effect of multiple stressors. An underlying theme of these challenges is that in addition to the question of what the chemical does to the biological receptor, we need to pay more and more attention to the question of how the biological receptor handles the chemicals, i.e. through which pathways the initial chemical-biological reaction takes. The interaction extends to adverse effects, phenotypic characteristics of the biological receptor as well as how it is modulated by adaptive or compensatory processes (Timbrell and Barile 2023).

Ecology seeks to understand the interactions between organisms and the environment to improve our ability to predict ecosystem changes and to identify general rules that explain the functioning of ecosystems (Loreau 2010). Significant problems in both freshwater and marine environments are reported as follows: (1) What are the processes that control the structure and functioning of aquatic ecosystems? (2) What ecological patterns are emerging at various spatio-temporal scales and what are their main drivers? (3) How aquatic organisms will respond to increasing anthropogenic pressures? (Martini et al. 2021a, b; Heino et al. 2015).

In the modern anthropogenic era, xenobiotic substances (dyes, pharmaceuticals and personal care products, endocrine active compounds, pesticides and polycyclic aromatic hydrocarbons) have a negative impact on biodiversity. The continuous development of industry and insufficiently successful waste management leads to increasing amounts of pollutants entering surface waters, especially xenobiotics. The main transfer pathways to aquatic ecosystems include industrial wastewater discharges, waste from domestic, industrial and agricultural sources. Therefore, there is a great need to identify the transfer and transformation pathways of xenobiotics in the environment, assess the impact of these substances on individual levels of the trophic chain and characterize their potential toxicity. Not only are xenobiotic compounds characterized by poor degradability, some can undergo partial decomposition, which can lead to the production of substances with higher toxic potential than the original compound. Xenobiotic compounds can also bioaccumulate in organisms, causing their concentrations in the tissues of animals to increase significantly as they translocate through levels of the trophic chain, a process called biomagnification (Dubey et al. 2014). Exposure of aquatic organisms to xenobiotics not only poses a constant threat, but also creates an urgent need to monitor the threats they pose and identify modern methods of remediation of contaminated ecosystems (Piwowarska and Kiedrzyńska 2022).

2.1.3 Pollution Transport in Aquatic Systems

In general, aquatic toxicology refers to the effects of a chemical on aquatic organisms, and these adverse effects are determined by organisms representing three trophic levels, namely (i) algae or plants representing "primary producers", (ii) invertebrates (e.g. crustaceans such as Daphnia spp.) representing "primary consumers/secondary producers", (iii) vertebrates (usually fish) representing "secondary consumers".

When pollutants are released into aquatic environments, they have direct (toxic) and indirect (secondary) effects on aquatic biota. Direct effects vary according to the intensity and duration of exposure to a toxicant. Species responses to pollutants are often studied to determine harmless levels of contamination (Long et al. 1995). These criteria are often derived from laboratory toxicity tests using model species in single toxicant exposures. Direct effects of toxicants typically reduce organism abundance (through increased mortality or reduced reproduction). Biota from a given habitat often exhibit a wide range of tolerance to certain toxicants (e.g. insecticides and herbicides targeting specific organisms in an interacting community). As a result, a toxicant may have lethal effects on some species but no observable effect on others. However, in indirect (or secondary) effects, pollutants can act on tolerant species through a number of ecological mechanisms. In laboratory-based toxicity tests, the effects of indirect pollutants cannot be detected in single species studies. Therefore, research at the population, community or ecosystem level is necessary (Fleeger et al. 2003).

The vast majority of aquatic toxicology studies are conducted in laboratories or experimental units where all variables can be controlled. It can also be in the form of nature studies, which are misperceived to be easier. The results of these studies have reached a level of reliability and simulation environments developed and designed for this purpose actually provide an easier working environment for researchers. Natural environment studies can provide a more realistic picture of the health status of exposed organisms. However, these studies are not conducted under controlled conditions and therefore rarely provide important information such as toxicokinetics, exposure duration and possible synergistic or antagonistic effects of other contaminants. Environmental approaches usually aim to determine which contaminants are present in water, sediment and fish organs, as well as the morphophysiological effects observed in contaminated environments (Paschoalini and Bazzoli 2021).

In an ever-changing environment, under increasing anthropogenic pressures, one of the biggest challenges aquatic organisms face is coping with chemical pollution. Thanks to the wonderful traits built into their very nature, these organisms can tolerate pollutants up to a certain level. The level of this tolerance is determined not only by the intrinsic characteristics of the organism but also by external factors. Surface waters can contain tens to hundreds of thousands of chemicals and numerous transformation products, including plant protection products, biocides, pharmaceuticals, surfactants, personal care products and many more (Brack et al. 2019). Moreover, chemical pollution often combines with hydromorphological changes and/or other stressors, such as eutrophication, oxygen loss, pH and temperature change, to cause complex environmental change in aquatic animals that requires the simultaneous activation of multiple biochemical and physiological control mechanisms (Tenji et al. 2020).

Another aspect that complicates research for aquatic toxicologists is the different routes of entry of pollutants into aquatic organisms. There are three main routes through which chemicals can enter the body in aquatic organisms: Gills, skin and gastrointestinal tract (Wassenaar et al. 2020). In the case of absorption through the gills and skin, the chemicals go directly into the bloodstream and many organs, while ingested chemicals first undergo intestinal metabolism followed by absorption and hepatic biotransformation (Pickford et al. 2003). As a result of biotransformation in aquatic organisms, new compounds that are much more toxic may emerge. The rapidly developing science of aquatic toxicology has now gone beyond the effects of chemicals and has begun to examine the toxic effects of these biotransformation products in aquatic organisms.

Once in the body, chemicals are distributed among various tissues. The distribution process depends on the properties of the absorbed compounds, which may have a strong affinity for different biomolecules—blood proteins, structural proteins, storage lipids or membrane lipids—and accumulate primarily in organs rich in these biomolecules (Da Cuña et al. 2020). The next stage is complex metabolism, leading to the transformation of the absorbed compounds into various metabolites. Metabolic processes are particularly intense in the cells of organs such as the liver, gills and kidneys (Maculewicz et al. 2020). The final stage of interaction of chemicals with the body is excretion (Arnot and Gobas 2006). Compounds entering the body can be excreted not only as derivatives formed during metabolic processes, but also as parent compounds. In addition, the process of pseudo-elimination should be considered, which involves a decrease in the concentration of chemical compounds through dilution in tissues (Segner 2015). All the processes described above are responsible for the phenomenon of bioaccumulation, which represents the accumulation of environmental chemicals in the tissues of organisms. Bioaccumulation includes bioconcentration, which involves the absorption of chemicals from the environment only through dermal surfaces, excluding the respiratory and digestive system, and also the biomagnification chain, a process tightly linked to trophic transfer, where the concentration usually increases at subsequent levels of food (Yarsan and Yipel 2013). All these processes can be mathematically described using the appropriate factors—BCF (bioconcentration factor), BAF (bioaccumulation factor) and BMF (biomagnification factor). BCF is calculated as the ratio of the chemical concentration in the organism and the steady-state chemical concentration in the surrounding environment (Arnot and Gobas 2006). In the case of BAF, the possibility of ingesting chemical compounds through the digestive tract is also taken into account. BMF is an indicator that concerns only dietary intake and is expressed as the ratio of the chemical concentration in the organism to the chemical concentration in its diet (Maculewicz et al. 2020). When all these stages are taken into consideration, it is understood how important it is for an aquatic toxicologist to master the biology of aquatic organisms in order to start toxicological studies. In addition to biology, a background in biochemistry and physiology is absolutely necessary for aquatic toxicologists.

Depending on the ability of the species to cope with chronic chemical stress, overall responses can range from gradual population decline (sometimes leading to species extinction) to the development of biochemical and physiological adaptations as mechanisms for the successful maintenance of homeostasis in the polluted environment (Whitehead et al. 2017; Tenji et al. 2020). Once chemicals or their biotransformation products enter the organism, they cause numerous responses at various levels of organization in almost all organ systems, often resulting in mild/severe side effects. However, situations associated with chemical exposure have led to the development of tolerance, with the most important physiological adaptive mechanisms related to metabolism and excretion of the target chemical, increases in antioxidant responses and changes in the life cycle of fish (Hamilton et al. 2017).

Worldwide, large pesticide loads are used to increase yields in terrestrial environments (according to FAO, 4.1 million tons worldwide in 2018 FAOSTAT 2021). This puts environmental pressure on terrestrial and aquatic life (Stanley et al. 2016). In particular, most herbicides, insecticides and fungicides are non-selective toxicants that interfere with the molecular and cellular function of both target and non-target species. Once released, pesticides reach aquatic ecosystems and interact with various targets in the cells of fish and other living organisms to exert their toxic effects (Uçar et al. 2020, 2021). Toxicity in fish can result from the metabolization process, which can produce toxic metabolites or derivatives of oxidizing molecules (Lushchak et al. 2018). Fish have well-developed biotransformation systems that convert lipophilic toxicants into more water-soluble metabolites to facilitate excretion (Santana et al. 2018). These systems are programs installed as the tolerance of aquatic organisms to pollutants. It is thanks to these features that they are able to cope with pollutants up to certain levels. Following absorption, pesticides move through the fish's circulatory system until they reach the site of metabolization (liver) and final excretion (kidney) (Maxwell 2020). As pesticides enter hepatocytes, the main site of xenobiotic metabolism, they undergo stage I (oxidation, reduction and hydrolysis reactions) and/or stage II (conjugation reactions) of the biotransformation process (Banerjee and Ramaiah 2020; Santana et al. 2022).

As the protagonists of aquatic ecosystems, fish are a very important nutritious food source all over the world. Regrettably, aquatic ecosystems are highly exposed to anthropogenic pollutants. This raises questions about the availability of safe, high-quality food sources. To improve fish welfare and monitoring programs, it is important to identify fish reproduction and development pathways in relation to ecosystem stressors. Xenobiotics accumulate in the reproductive and developmental systems of the organism, causing toxicity. An advanced technique, such as the proteomics approach, can solve the current problem in species of economic importance in aquaculture as well as in the wild to assess the developmental biology, physiological status and impact of environmental toxicity in fish. Recent advances in the mass spectrometry approach of proteomics methods could lead to new horizons in the field of biology. Identified proteins can map important possible mechanisms of toxicity of xenobiotics in ecosystems (Pinto et al. 2023).

Due to global changes, aquatic environments are exposed to environmental stressors such as global warming, pollution or emerging pathogens (Lange et al. 2018). The targets to be achieved in this direction are: (I) What are the consequences of multiple environmental stressors on fish at molecular, physiological and whole organism levels? (II) How stress responses are integrated across levels of biological organization, especially under multiple stressors? (III) Why some populations are able to cope with environmental stressors while others are not? (IV) What are the long-term consequences of stressors over generations and the evolutionary mechanisms involved? To address these questions, multiple research groups have been formed to address the molecular and physiological pathways involved in the stress response, with an increase in the literature focusing on fish in recent years (Uçar et al. 2020, 2021, 2022, 2023, Alak et al. 2017, 2018, 2019, 2020, 2021, 2022; Atamanalp et al. 2021, 2022, 2023; Parlak et al. 2022). Based on the existing literature, four promising ways for future research can be suggested. These are; (1) exposure to multiple stressors leads to complex interactive effects (synergism or antagonism) at biological scales that should be considered in future studies, (2) such effects may result from shifts from a "metabolic compensation strategy" to a "metabolic protection strategy" when a certain threshold or combination of stressor(s) is reached, (3) how this might affect the population, and potentially why some populations are resilient to problems while others decrease, (4) then considering evidence of inter-population variability shown in fish stress responses and how the intergenerational effects of stressors can have strong consequences for the evolutionary trajectories of natural populations and in some cases lead to extinction or local adaptation. We believe that developing such promising ways of future research will significantly increase our understanding of fish response to current and future stressors in the context of global changes (Petitjean et al. 2019).

2.2 Conclusion

Human actions can cause environmental damage specific to sustainability. The use of synthetic chemicals as pesticides, the abundance of pharmaceuticals and personal care products in wastewater, industrial activities using paints or other organic compounds that can persist in nature for long periods of time are some examples of pollutant sources. This is a global problem that threatens the survival of various forms of life on the planet, including humans, and therefore the ecosystem as a whole.

Ecotoxicological studies have increased significantly in recent years following the exponential growth in the production and use of chemicals in agriculture, pharmaceuticals and various industrial sectors. This has led to an increasing release of toxic pollutants into waters globally. Inland waters in particular are among the most threatened habitats worldwide due to pollution. Their protection should be part of the priorities of environmental protection and management policies, given both the character of their biocenoses and their essential properties. Consequently, monitoring activities of water quality and environmental status are of great importance. In this respect, ecotoxicological studies play a prominent role in assessing pollutant impacts on aquatic habitats/human health.

Freshwater ecosystems play a magical role in supporting life processes in living organisms. However, increasing population, urbanization, industrialization have put a serious pressure on freshwater ecosystems, changing the quality scenario of freshwater in large quantities. While these changes can be clearly assessed in water resources with past records, the lack of previous studies in undeveloped countries makes global assessment impossible. The fact that these countries stay behind in industrialization has given them a chance in this regard. In developed countries, since industrialization started in the early years, the history of studies to examine the effects of pollution also dates back to ancient times. The increase in information sharing with globalization is important for humanity and allows the measures to be taken to be worldwide. Undeveloped countries have the chance to take measures at earlier stages by benefiting from the negative experiences of industrialized countries.

Water pollution not only degrades the quality of freshwater, but also poses harmful risks to the entire environment. Due to the complex nature of chemicals in freshwater ecosystems, they are not easily neutralized and have a high potential to remain intact in any environment. Today, these substances are continuously added to freshwater ecosystems through the discharge of untreated domestic, industrial and agricultural wastewater. Most of these substances accumulate in bottom sediments and very small concentrations in the form of organic/inorganic components remain suspended or in solution in the water column of the freshwater ecosystem. The presence of such pollutants in the freshwater ecosystem has long-term effects on aquatic biota. Therefore, the focus should be on regular monitoring of the quality of the freshwater ecosystem and treatment of wastewater before it is discharged into the freshwater ecosystem.

Acknowledgements Many thanks to my colleagues from Fisheries Faculty who supported me in the research for their help Atatürk University for their support in my academic studies.

References

Alak G, Parlak V, Ucar A, Yeltekin AC, Ozgeris FB, Caglar O, ... Turkez H (2020) Oxidative and DNA damage potential of colemanite on zebrafish: Brain, liver and blood. Turk. J. Fish. Aquat. Sci 20(8):593–602

Alak G, Uçar A, Parlak V, Atamanalp M (2022) Identification, characterisation of microplastic and their effects on aquatic organisms. Chem Ecol 38(10):967–987

Alak G, Ucar A, Parlak V, Yeltekin AÇ, Özgeriş FB, Atamanalp M, Türkez H (2021) Antioxidant potential of ulexite in zebrafish brain: assessment of oxidative DNA damage, apoptosis, and response of antioxidant defense system. Biol Trace Elem Res 199:1092–1099

Alak G, Ucar A, Yeltekin AÇ, Çomaklı S, Parlak V, Taş IH, ... Türkez H (2018) Neuroprotective effects of dietary borax in the brain tissue of rainbow trout (Oncorhynchus mykiss) exposed to copper-induced toxicity. Fish Physiol Biochem 44:1409–1420

Alak G, Yeltekin AÇ, Tas IH, Ucar A, Parlak V, Topal A, ... Atamanalp M (2017) Investigation of 8-OHdG, CYP1A, HSP70 and transcriptional analyses of antioxidant defence system in liver tissues of rainbow trout exposed to eprinomectin. Fish Shellfish Immunol 65:136–144

Alak G, Yeltekin AÇ, Uçar A, Parlak V, Türkez H, Atamanalp M (2019) Borax alleviates copper-induced renal injury via inhibiting the DNA damage and apoptosis in rainbow trout. Biol Trace Elem Res 191:495–501

Arnot JA, Gobas FA (2006) A review of bioconcentration factor (BCF) and bioaccumulation factor (BAF) assessments for organic chemicals in aquatic organisms. Environ Rev 14(4):257–297. https://doi.org/10.1139/a06-005

Atamanalp M, Kırıcı M, Köktürk M, Kırıcı M, Kocaman EM, Ucar A, ... Alak G (2023) Polyethylene exposure in rainbow trout; suppresses growth and may act as a promoting agent in tissue-based oxidative response, DNA damage and apoptosis. Process Saf Environ Prot 174:960–970

Atamanalp M, Köktürk M, Uçar A, Duyar HA, Özdemir S, Parlak V, ... Alak G (2021) Microplastics in tissues (brain, gill, muscle and gastrointestinal) of Mullus barbatus and Alosa immaculata. Arch Environ Contam Toxicol 81:460–469

Atamanalp M, Türkez H, Yeltekin AÇ, Özgeriş FB, Ucar A, Çağlar Ö, ... Alak G (2022) Borax relieved the acrylamide-induced hematotoxic, hepatotoxic, immunotoxic and genotoxic damages in rainbow trout by regulating apoptosis and Nrf2 signaling pathway. Comp biochem physiol C Toxicol pharmacol 259:109396

Banerjee A, Ramaiah SK (2020) Hepatic toxicology. In: An introduction to interdisciplinary toxicology. Academic Press, pp 153–162. https://doi.org/10.1016/B978-0-12-813602-7.00012-0

Bashir I, Lone FA, Bhat RA, Mir SA, Dar ZA, Dar SA (2020) Concerns and threats of contamination on aquatic ecosystems. Bioremediation and biotechnology: sustainable approaches to pollution degradation 1–26. https://doi.org/10.1007/978-3-030-35691-0_1

Brack W, Aissa SA, Backhaus T, Dulio V, Escher BI, Faust M, Altenburger R (2019) Effect-based methods are key. The European Collaborative Project SOLUTIONS recommends integrating effect-based methods for diagnosis and monitoring of water quality. Environ Sci Eur 31(1):1–6. https://doi.org/10.1186/s12302-019-0192-2

Brady SP, Richardson JL, Kunz BK (2017) Incorporating evolutionary insights to improve ecotoxicology for freshwater species. Evol Appl 10(8):829–838. https://doi.org/10.1111/eva.12507

Da Cuña RH, Lo Nostro FL, Shimabukuro V, Ondarza PM, Miglioranza KSB (2020) Bioaccumulation and distribution behavior of endosulfan on a cichlid fish: differences between exposure to the active ingredient and a commercial formulation. Environ Toxicol Chem 39(3):604–611. https://doi.org/10.1002/etc.4643

Davis AP, Grondin CJ, Lennon-Hopkins K, Saraceni-Richards C, Sciaky D, King BL, ... Mattingly CJ (2015) The Comparative Toxicogenomics Database's 10th year anniversary: update 2015. Nucleic Acids Res 43(D1):D914–D920

Don-Pedro KN (1990) Insecticidal activity of fatty acid constituents of fixed vegetable oils against Callosobruchus maculatus (F.) on cowpea. Pesticide Sci 30(3):295–302. https://doi.org/10.1002/ps.2780300306

Dubey KK, Kumar P, Singh PK, Shukla P (2014) Exploring prospects of monooxygenase-based biocatalysts in xenobiotics. In: Microbial biodegradation and bioremediation. Elsevier, pp 577–614. https://doi.org/10.1016/B978-0-12-800021-2.00026-1

Ekubo AJ, Abowel JFN (2011) Aspects of aquatic pollution in Nigeria. Res J Environ Earth Sci 3:673–693

Fleeger JW, Carman KR, Nisbet RM (2003) Indirect effects of contaminants in aquatic ecosystems. Sci Total Environ 317(1–3):207–233. https://doi.org/10.1016/S0048-9697(03)00141-4

Forster J, Hirst AG, Atkinson D (2012) Warming-induced reductions in body size are greater in aquatic than terrestrial species. Proc Natl Acad Sci 109(47):19310–19314. https://doi.org/10.1073/pnas.1210460109

Fricke R, Eschmeyer WN, Van der Laan R (2018) Catalog of fishes: genera, species, references. California Academy of Sciences, San Francisco, CA, USA http://researcharchive.calacademy.org/research/ichthyology/catalog/fishcatmain.asp

Gupta PK (2016) Chapter 3—Classification of poisons/toxicants. In Gupta PK (ed) Fundamentals of toxicology. Academic Press, pp 19–21 (2016)

Hamilton PB, Rolshausen G, Uren Webster TM, Tyler CR (2017) Adaptive capabilities and fitness consequences associated with pollution exposure in fish. Philosop Trans Roy Soc B: Biol Sci 372(1712):20160042. https://doi.org/10.1098/rstb.2016.0042

Hampel M, Blasco J, Segner H (2015) Molecular and cellular effects of contamination in aquatic ecosystems. Environ Sci Pollut Res 22:17261–17266. https://doi.org/10.1007/s11356-015-5565-5

Heino J, Soininen J, Alahuhta J, Lappalainen J, Virtanen R (2015) A comparative analysis of metacommunity types in the freshwater realm. Ecol Evol 5(7):1525–1537. https://doi.org/10.1002/ece3.1460

Hellou J (2011) Behavioural ecotoxicology, an "early warning" signal to assess environmental quality. Environ Sci Pollut Res 18:1–11. https://doi.org/10.1007/s11356-010-0367-2

Hughes RM (2015) Recreational fisheries in the USA: economics, management strategies, and ecological threats. Fish Sci 81(1):1–9. https://doi.org/10.1007/s12562-014-0815-x

Ibor OR, Adeogun AO, Regoli F, Arukwe A (2019) Xenobiotic biotransformation, oxidative stress and obesogenic molecular biomarker responses in Tilapia guineensis from Eleyele Lake, Nigeria. Ecotoxicol Environ Saf 169:255–265. https://doi.org/10.1016/j.ecoenv.2018.11.021

Klaminder J, Jonsson M, Fick J, Sundelin A, Brodin T (2014) The conceptual imperfection of aquatic risk assessment tests: highlighting the need for tests designed to detect therapeutic effects of pharmaceutical contaminants. Environ Res Lett 9(8):084003. https://doi.org/10.1088/1748-9326/9/8/084003

Lange K, Bruder A, Matthaei CD, Brodersen J, Paterson RA (2018) Multiple-stressor effects on freshwater fish: Importance of taxonomy and life stage. Fish Fish 19(6):974–983. https://doi.org/10.1111/faf.12305

Long ER, MacDonald DD, Smith SL, Calder FD (1995) Incidence of adverse biological effects within ranges of chemical concentrations in marine and estuarine sediments. Environ Manage 19:81–97. https://doi.org/10.1007/BF02472006

Loreau M (2010) Linking biodiversity and ecosystems: towards a unifying ecological theory. Philos Trans Roy Soc B: Biol Sci 365(1537):49–60. https://doi.org/10.1098/rstb.2009.0155

Lushchak VI (2011) Environmentally induced oxidative stress in aquatic animals. Aquat Toxicol 101(1):13–30. https://doi.org/10.1016/j.aquatox.2010.10.006

Lushchak VI, Matviishyn TM, Husak VV, Storey JM, Storey KB (2018) Pesticide toxicity: a mechanistic approach. EXCLI j 17:1101

Maculewicz J, Świacka K, Kowalska D, Stepnowski P, Stolte S, Dołżonek J (2020) In vitro methods for predicting the bioconcentration of xenobiotics in aquatic organisms. Sci Total Environ 739:140261. https://doi.org/10.1016/j.scitotenv.2020.140261

Martin O, Scholze M, Ermler S, McPhie J, Bopp SK, Kienzler A, Kortenkamp A (2021) Ten years of research on synergisms and antagonisms in chemical mixtures: a systematic review and quantitative reappraisal of mixture studies. Environ Int 146:106206. https://doi.org/10.1016/j.envint.2020.106206

Martini S, Larras F, Boyé A, Faure E, Aberle N, Archambault P, Ayata SD (2021) Functional trait-based approaches as a common framework for aquatic ecologists. Limnol Oceanogr 66(3):965–994. https://doi.org/10.1002/lno.11655

Maxwell L (2020) Absorption, distribution, and excretion in complex organisms. In An introduction to interdisciplinary toxicology. Academic Press, pp 17–29 https://doi.org/10.1016/B978-0-12-813602-7.00002-8

Namba H, Iwasaki Y, Heino J, Matsuda H (2020) What to survey? A systematic review of the choice of biological groups in assessing ecological impacts of metals in running waters. Environ Toxicol Chem 39(10):1964–1972. https://doi.org/10.1002/etc.4810

Parlak V, Ozgeris B, Ucar A, Cilingir Yeltekin A, Ozgeris FB, Cağlar O, ... Atamanalp M (2022) Hematotoxic, oxidative and genotoxic damage in rainbow trout (Oncorhynchus mykiss) after exposure to 3-benzoylpyridine. Toxicol Mech Methods 32(7):501–509

Paschoalini AL, Bazzoli N (2021) Heavy metals affecting Neotropical freshwater fish: a review of the last 10 years of research. Aquat Toxicol 237:105906. https://doi.org/10.1016/j.aquatox.2021.105906

Petitjean Q, Jean S, Gandar A, Côte J, Laffaille P, Jacquin L (2019) Stress responses in fish: from molecular to evolutionary processes. Sci Total Environ 684:371–380. https://doi.org/10.1016/j.scitotenv.2019.05.357

Pickford KA, Thomas-Jones RE, Wheals B, Tyler CR, Sumpter JP (2003) Route of exposure affects the oestrogenic response of fish to 4-tert-nonylphenol. Aquat Toxicol 65(3):267–279. https://doi.org/10.1016/S0166-445X(03)00149-8

Pinto N, Yashwanth BS, Radhakrishnan G, Sathiyanarayanan A (2023) A Proteomic approach to studying the effects of xenobiotics on aquatic living organisms. In: Xenobiotics in aquatic animals: reproductive and developmental impacts. Springer Nature Singapore, Singapore, pp 233–250. https://doi.org/10.1007/978-981-99-1214-8_10

Piwowarska D, Kiedrzyńska E (2022) Xenobiotics as a contemporary threat to surface waters. Ecohydrol Hydrobiol 22(2):337–354. https://doi.org/10.1016/j.ecohyd.2021.09.003

Rand GM (Ed.) (1995) Fundamentals of aquatic toxicology: effects, environmental fate and risk assessment. CRC press

Reid AJ, Carlson AK, Creed IF, Eliason EJ, Gell PA, Johnson PTJ, 189 Kidd KA, acCormack TJ, Olden JD, Ormerod SJ, Smol JP, Taylor WW, Tockner K, Vermaire JC, Dudgeon D, Cooke SJ

(2018) Emerging threats and persistent conservation challenges for freshwater biodiversity. Biol Rev 94:849–873. https://doi.org/10.1111/brv.12480
Rudén C, Adams J, Ågerstrand M, Brock TC, Poulsen V, Schlekat CE, Henry TR (2017) Assessing the relevance of ecotoxicological studies for regulatory decision making. Integr Environ Assess Manag 13(4):652–663. https://doi.org/10.1002/ieam.1846
Samantha JJ, Meredith GL (2020) Chapter 26 - Environmental toxicology: aquatic. In: Wexler P (ed) Information resources in toxicology (4th edn). Academic Press, pp 263–278
Santana MS, de Melo GD, Sandrini-Neto L, Di Domenico M, Prodocimo MM (2022) A meta-analytic review of fish antioxidant defense and biotransformation systems following pesticide exposure. Chemosphere 291:132730. https://doi.org/10.1016/j.chemosphere.2021.132730
Santana MS, Sandrini-Neto L, Neto FF, Ribeiro CAO, Di Domenico M, Prodocimo MM (2018) Biomarker responses in fish exposed to polycyclic aromatic hydrocarbons (PAHs): systematic review and meta-analysis. Environ Pollut 242:449–461. https://doi.org/10.1016/j.envpol.2018.07.004
Segner H (2015) In vitro methodologies in ecotoxicological hazard assessment: the case of bioaccumulation testing for fish. Altern Lab Anim 43(2):P14–P16. https://doi.org/10.1177/026119291504300209
Stanley J, Preetha G, Stanley J (2016) Pesticide toxicity to non-target organisms, vol 502. Springer, Berlin, Germany. https://doi.org/10.1007/s11356-020-10245-w
Strayer DL, Dudgeon D (2010) Freshwater biodiversity conservation: recent progress and future challenges. J N Am Benthol Soc 29(1):344–358. https://doi.org/10.1899/08-171.1
Tenji D, Micic B, Sipos S, Miljanovic B, Teodorovic I, Kaisarevic S (2020) Fish biomarkers from a different perspective: evidence of adaptive strategy of Abramis brama (L.) to chemical stress. Environ Sci Eur 32:1–15
Thoré ES, Philippe C, Brendonck L, Pinceel T (2021) Towards improved fish tests in ecotoxicology-efficient chronic and multi-generational testing with the killifish Nothobranchius furzeri. Chemosphere 273:129697. https://doi.org/10.1016/j.chemosphere.2021.129697
Timbrell J, Barile FA (2023) Introduction to toxicology. CRC Press
Ucar A, Ozgeris FB, Parlak V, Yeltekin AC, Turkez H, Alak G, Atamanalp M (2022) Ulexite modulates the neurotoxicological outcomes of acetylferrocene-exposed rainbow trout. Environ Mol Mutagen 63(6):286–295
Uçar A, Parlak V, Çilingir Yeltekin A, Özgeriş FB, Çağlar Ö, Türkez H, ... Atamanalp M (2021) Assesment of hematotoxic, oxidative and genotoxic damage potentials of fipronil in rainbow trout Oncorhynchus mykiss, Walbaum. Toxicol Mech Methods 31(1):73–80
Uçar A, Parlak V, Özgeriş FB, Yeltekin AÇ, Alak G, Atamanalp M (2020) Determination of Fipronil toxicity by different biomarkers in gill and liver tissue of rainbow trout (Oncorhynchus mykiss). In Vitro Cell Dev Biol Anim 56:543–549
Ucar A, Yeltekin AÇ, Köktürk M, Calimli MH, Nas MS, Parlak V, ... Atamanalp M (2023) Has PdCu@ GO effect on oxidant/antioxidant balance? Using zebrafish embryos and larvae as a model. Chem Biol Interact 378:110484
Wassenaar PN, Verbruggen EM, Cieraad E, Peijnenburg WJ, Vijver MG (2020) Variability in fish bioconcentration factors: Influences of study design and consequences for regulation. Chemosphere 239:124731. https://doi.org/10.1016/j.chemosphere.2019.124731
Whitehead A, Clark BW, Reid NM, Hahn ME, Nacci D (2017) When evolution is the solution to pollution: key principles, and lessons from rapid repeated adaptation of killifish (Fundulus heteroclitus) populations. Evol Appl 10(8):762–783. https://doi.org/10.1111/eva.12470
Yarsan E, Yipel M (2013) The important terms of marine pollution "biomarkers and biomonitoring, bioaccumulation, bioconcentration, biomagnification. J Mol Biomark Diagn S 1(2). https://doi.org/10.4172/2155-9929.S1-003

Chapter 3
Model Organisms Used in Aquatic Toxicology

Veysel Parlak

Abstract Much of our knowledge about heredity, development, physiology, cellular and molecular processes in living organisms is derived from the study of model organisms. Despite the enormous biodiversity of the Earth, a common basis of mechanisms for all living things can be deduced by studying a few life forms selected for their suitability for experimental studies. For example, basic metabolisms can be elucidated with several model organisms, including *E. coli*, yeast, *C. elegans*, *Drosophila*, *Xenopus*, zebrafish, mouse, maize and Arabidopsis. These model organisms have become the metabolisms of choice for comprehensive genome studies (transcriptomics and proteomics) due to their wide range of applications and important properties. Gregor Mendel is known as the first person to experimentally select an organism to solve questions of universal inheritance. Mendel's work has guided the selection of model organisms in subsequent times. Since Mendel's pea experiments, the revolutionary knowledge gained has been largely based on model organism studies. Thanks to model organisms, many of the molecular mechanisms elucidated at the cellular level have revealed the kinship of life on earth at a level invisible to the naked eye. In today's biology, the 'model organism' helps us to understand how living organisms work. The main factor in the selection of species for model organisms is the focus of the studies. Model organisms are organisms that are produced under in vitro conditions and are easy to maintain production, have a short generation interval, have experimental advantages and are used in the investigation of biological phenomena. Model organisms, which have high homology but small genome size compared to the human genome, can be used in all kinds of experiments that cannot be studied on humans. These organisms are easy to produce and maintain in the laboratory and offer considerable experimental advantages. They are also inexpensive to produce and their use is not ethically compromised. These organisms are usually selected from widely available and easily cultivated species. A good model organism is one that is genetically defined, capable of rapid reproduction, has sensitive responses and

V. Parlak (✉)
Department of Basic Sciences, Faculty of Fisheries, Atatürk University, Erzurum, Türkiye
e-mail: veysel.parlak@atauni.edu.tr

M. Atamanalp et al., *Aquatic Toxicology in Freshwater*, Springer Water,
https://doi.org/10.1007/978-3-031-56669-1_3

can produce reproducible results in experimental studies using genetically similar individuals.

Keywords Aquatic toxicology · Model organisms · Fish · Rainbow trout · Zebrafish

List of Abbreviation

DNA Deoxyribonucleic acid

3.1 Introduction

As in traditional toxicology, model organisms play a key role in environmental monitoring research. In researches, model organisms are used to identify chemicals present in the environment as well as to investigate the fate and effects of these chemicals in the ecosystem. For research and regulatory applications, teleost fish have been studied more than any other group (Ligo et al. 2007; Alejo and Tafalla 2011; Forsgren et al. 2013; Andreassen and Høyheim 2017; Luo et al. 2019; Stosiket al. 2021; Nati et al. 2023) and many species have been included in both acute and chronic toxicity tests.

Fish species considered as model organisms include rainbow trout and other trout species (LaFrentz et al. 2012; Evenhuis and LaFrentz 2016; Eriksson-Kallio et al. 2020), zebrafish (Peterson et al. 2013; Patrikainen et al. 2017; Parlak 2018; Bradford et al. 2022); American flagfish (Smith et al. 1991; Orrego et al. 2011), fat minnows (*Pimephales promolus*) (Jensen et al. 2001; Valenti et al. 2009), killifish/mummichogs (Marshall et al. 2002; Judith and Candelmo 2012), spiny fish (Friedma 2010; Near et al. 2013; Ghezelayagh et al. 2022) and Japanese medaka (Kawakami et al. 2000; Shima and Mitani 2004; Nassef et al. 2010).

Freshwater invertebrates include cladocerans (Jeppesen et al. 2001; Sommer and Sommer 2006; Leppänen 2018), rotifers (Snell and Janssen 1995; Wallace 2002; Gladyshev and Meselson 2008; Dahms et al. 2011; García-Roger et al. 2019; Park et al. 2020; Gilbert 2022), crayfish (Larson and Olden 2016; Huang et al. 2019; Chen et al. 2020; Jiang and Cao 2021) and bivalve molluscs (Gomez-Chiarri et al. 2015; Wang and Lu 2017; Bonfim et al. 2019) were used.

Microalgae (Santos-Ballardo et al. 2015; Xiong et al. 2017; Metsoviti et al. 2019), protozoa (Montagnes et al. 2012; Newbold et al. 2015; Weidner et al. 2017; Firkins et al. 2020) and amphibians (do Amaral et al. 2019; Muths et al. 2020; Bodensteiner et al. 2021) are also used in toxicology applications. Marine invertebrates include polychaetes (Bruschetti 2019; Calmao et al. 2023), molluscs (Matthiessen 2008; Verdes et al. 2016), crustaceans (copepods, mussels, ten-legged) (Boese et al. 1997;

Rainbow and Luoma 2011; Romano and Zeng 2013; Parrella et al. 2014; D'Costa 2022), and echinoderms (Bay et al. 1993; Nipper et al. 1993; Nobre et al. 2015; Genevière et al. 2020; Anagha et al. 2022) have been extensively studied. It is also known that artemia and urchin embryos/larvae and bacteria have been used in acute and chronic toxicity tests (Bergam et al. 2017; da Silva and de Souza Abessa 2019; Prabhakar et al. 2021).

Model organisms are selected for their genetic and biological similarity to the human genome and help in the interpretation of human genomic sequence. In addition, factors such as the short intergenerational time between generations and easy monitoring of embryonic development are also effective in the selection of a living organism as a model organism. Genome mapping of model organisms allows understanding of gene regulation and the mechanism of human genetic diseases and developmental, physiological and biological processes (Bonini and Berger 2017). The interspecific genetic similarities that exist in all organisms mean that the study of one organism can be used as a source of data for other species (Miller et al. 2004). Model organisms are used to learn about other species (including humans) that are difficult to study.

When selecting model organisms to be used in studies, factors such as high homology with the human genome,—short intergenerational time,—easy monitoring of embryonic development and intervention,—easy culturing,—suitable for experimental manipulations,—suitable for genetic analysis and not causing much ethical problems should be considered.

There are 3 different types of model organisms;

1. Genetic Model Organisms
2. Experimental Model Organisms
3. Genomic Model Organisms

1. **Genetic Model Organisms**

These are suitable organisms for genetic analysis. These organisms reproduce very quickly and produce many generations in a short time. Usually a lot of mutant individuals are obtained. Genetic maps can be made. Examples of these species are *Saccharomyces cerevisiae, Drosophila melanogaster* and *Caenorhabditis elegans.*

2. **Experimental Model Organisms**

These organisms are not used in genetic research because they have long generation times and a poor genetic map. However, hardy embryos produced by species such as *Gallus gallus* (chicken) and *Xenopus laevis* (a frog species) are used as model organisms in developmental biology.

3. **Genomic Model Organisms**

Such organisms are chosen as model organisms regardless of their genetic or experimental advantages and disadvantages. Because they are at a crucial point in evolutionary development or because some quality of their genome makes them ideal to study. *Fugu rubripes*, for example, have genes similar to those of humans, but smaller genomes. This difference in size is due to DNA repeats and introns in the human genome. Similarly, the zebrafish (*Danio rerio*) is the only genetically accessible model, especially for appendage and heart regeneration.

A very important point about model organisms is that their genes are closely related to human genes. More than 60% of the human disease genes identified so far are surprisingly similar to those of fruit flies and nematodes. Genes that influence advanced evolutionary development, such as our immune system, are less similar to those in simple animals.

For such genes, we need a model organism closer to us, such as a mouse. The organization of mouse genomes is similar to the organization of human genomes. Mice are used to prove disease models by mimicking defective genes that appear in humans. They can also be used to test the efficacy of new drugs. Although model organisms are mainly used for biomedical studies, in the second step they are used as biomarkers for the study of the degradation of natural ecosystems or their components.

In recent years, macro- and micro-pollutants in the aquatic environment have caused an increase in environmental concern and have become an important issue that has been studied worldwide. It is very important to detect and prevent pollution from hazardous substances in and around water and to determine the physiological and biochemical effects that may occur in aquatic organisms. In the simplest case, aquatic toxicology studies try to evaluate the mechanisms that toxic substances may have at the organismal or cellular level. These studies are conducted in laboratory exposures using a single organism and a single pure toxicant. In practice, exposure time is usually limited. In line with this information, model organisms used in toxicology tests in aquatic environment are given below.

***Daphnia magna* (Water flea)**: Daphnia magna is a fast reproducing and easily cultivated species found in freshwater lakes and rivers. It is frequently used to assess toxic effects and is a common model organism, especially for investigating the effects of aquatic pesticides. Because they are well characterized, have a rapid parthenogenetic reproductive cycle and show sensitivity to a range of environmental xenobiotics, they are well established and widely used for freshwater toxicity testing (Orsini et al. 2012; Aalizadeh et al. 2017; Lee et al. 2019; Schirmer et al. 2022).

***Cyprinodon variegatus* (*Sheepshead minnow*)**: The sheepshead minnow is a small marine fish that lives in coastal areas. It is a popular model organism used to investigate environmental toxic effects (Hemmer et al. 2001; Raimondo et al. 2009, 2016; Winder et al. 2012; Kirby et al. 2020; Gaillard et al. 2023).

Ceriodaphnia dubia: A species of water flea found in aquatic environments. It is mainly used to assess the toxic effects of industrial waste and environmental pollution (Henry and Black 2007; Nys et al. 2014; Mount et al. 2019; Connors et al. 2022).

Chironomus sp. (Mayflies): Chironomus species are larval stage model organisms commonly found in aquatic ecosystems and used in toxicological studies. Used to assess environmental toxic effects and determine water quality (Clare et al. 2011; Wolters et al. 2018; dos Reis Oliveira et al. 2020; Carey et al. 2023).

These are just a few examples and the list of model organisms used in aquatic toxicology is much wider. Each organism has different advantages, uses and sensitivities. Researchers select model organisms according to their specific needs and study objectives.

***Danio rerio* (Zebrafish)**: Zebrafish is a species that lives in both fresh and salt water, with fast growth and reproduction. It is widely used in the study of toxic effects at the embryo and adult stages. It is a good organism for vertebrate development because it has the best features of all other models. As in frogs, the embryo of the zebrafish develops in the external environment and all its stages can be monitored and modified. The development of the embryo is rapid, its organization simple and transparent. It is suitable for genetic analysis. They produce 200 eggs per week. Gene transfer technologies to zebrafish are very advanced. The genetic map of zebrafish, useful for comparative mapping of human genes, has been mapped thanks to these technologies. Many human disease and developmental genes are found in the zebrafish genome. It has 1.7 billion base pairs. The mutant zebrafish produced are a suitable model for human diseases. It is used in drug trials. Models include Alzheimer's disease, congenital heart disease, polycystic kidney disease and cancer (Gioacchini et al. 2010; Jaja-Chimedza et al. 2012; Babayigit et al. 2016; Gulías et al. 2019; Canedo et al. 2022; Kuroda et al. 2023).

Japanese puffer, Tiger puffer, or torafugu (*Fugu rubripes*): It is used as a model organism because it has the smallest genome among vertebrates. It is ¼ the size of the zebrafish genome and 1/10 the size of the human genome. Although the genome of the fish is short, it contains a similar number of genes to those found in the human genome. However, the genes of *F. rubripes* are organized differently and contain fewer introns. The small size of their genomes makes it easier to detect and analyze genes (Kamijo et al. 2011; Wang et al. 2015; Lin et al. 2020; Zhang et al. 2023).

The genome sequence of *F. rubripes* was published in 2002 and is the second vertebrate genome to be completed after the human genome. The genome of *F. rubripes* was assembled using shotgun sequencing and more than 1000 previously unidentified human genes were uncovered with the help of comparative genomics.

Frog: *Xenopus leavis* is successful model organisms in many ways. They are not suitable for genetic studies but are good experimental models of vertebrate development. They are not suitable for genetic studies because the offspring produced take too long to develop (4 months to reach sexual maturity).

X. tropicalis is a species close to *X. laevis*. It is suitable for genetic studies because it is diploid and has a short developmental period. Since the development of the frog embryo is in water, all stages can be monitored (Jensen et al. 2012; Junker and Rapoport 2015; Stachurska and Sarna 2019; El Fessi et al. 2022).

Medaka fish (*Oryzias latipes*): Medaka is a freshwater fish, with 3–4 cm size, spawns and egg development occurs externally. Both eggs and embryos are transparent. Medaka can live in a wide range of salinity and temperature (10–40 °C), are easy to breed and resistant to disease (Jung et al. 2014; Uemura et al. 2015; Fang et al. 2018; Imangali et al. 2021; Lu et al. 2023).

Rainbow trout (*Oncorhynchus mykiss*): Rainbow trout, a member of the Salmonidae family, is the most widely farmed freshwater fish species worldwide. Rainbow trout is a model organism used to identify adverse effects of environmental chemicals and to develop strategies to prevent them (LaFrentz et al. 2012; Murashita et al. 2014; Minghetti et al. 2017; Shomorin et al. 2019; Blouin et al. 2021; Atamanalp et al. 2022; Ucar et al. 2023).

Nile tilapia (*Oreochromis niloticus*): Nile tilapia is a species belonging to the Cichlidae family, endemic to Africa and living in freshwater. Nile tilapia is preferred as a model organism in experimental studies due to its economic importance, widespread cultivation, easy availability and low chromosome number ($2n = 44$) (Yanbo et al. 2006; Abdel-Warith et al. 2011; Naiel et al. 2020; Raza et al. 2023).

Sea urchin (*Paracentrotus lividus*): The sea urchin, an edible species, belongs to the family Echinoidea of the phylum Echinodermata. It is naturally distributed along the European coasts of the Mediterranean and North Atlantic. The sea urchin embryo has been used as a model organism for long-term developmental biology studies (Šiller et al. 2013; Salvo et al. 2016; Oliviero et al. 2019; Morroni et al. 2022; Ignoto et al. 2023).

Freshwater ciliates: Freshwater ciliates are one of the most important groups of protista. They have a ciliated structure and can reach up to 2 mm in length. Similar to Daphnia, most planktonic ciliates are algivorous. They live in natural lakes and reproduce asexually (Buonanno et al. 2012; Abraham et al. 2017; Vilas–Boas et al. 2020; Moeller and Johnson 2023).

Platyhelminthes (*Macrostomum lignano*): They are flatworms in the family *Macrostomorpha.* They are hermaphrodites and adults can grow up to 1.7 mm in length. This genus includes more than 100 marines, freshwater and brackish water species with wide distributions. This new species can be easily cultured under laboratory conditions and used in developmental studies (Tietje et al. 2013; Zadesenets et al. 2017; Bickmeyer et al. 2020; Ma et al. 2023) (Table 3.1).

3.2 Some Advantages of Using Model Organisms in Aquatic Toxicology

3.2.1 Experimental Control

Model organisms are used for experiments that can be better controlled and monitored in a laboratory setting. This is important for reproducibility of experimental conditions and reliability of results.

Table 3.1 The use of aquatic model organisms in toxicology studies

Aquatic model organisms	The most intensive purposes of use	Web of science search
Rainbow trout (*Oncorhynchus mykiss*)	Bioaccummulation, Macroinvertabrates, Ketotestoserone, *Vibro cholrea*, infectious, Microsatellitess, Digenea, Meat quality, Fisheries, Growth hormone	30,403 quantity article (09/07/2023)
Water flea (*Daphnia* pulex)	Marine Biology, Herbicides, pescides and Ground Poisoning, Nanoparticles, Phylogenetics Water treatment, Entomology, Ecology	18,665 quantity article (09/07/2023)
Zebrafish (*Danio reiro*)	Molecular and Cell Biology, Cancer and development, Toxicology, Marine Biology, Nanoparticles, Genetics, Nuroscience, Ecology, Immunology, transplantation and Microinjection Studies	16,783 quantity article (09/07/2023)
Tilapia (*O. niloticus*)	Phylogenetics and Genomics, Water treatment, Entomology, Protein structure, Contamination	14,086 quantity article (09/07/2023)
Medaka fish (*Oryzias latipes*)	Herbicides, pescides and Ground Poisoning, Marine Biology, Cancer and Development, Chromosome disorders, Dna Damage, Reproductive biology	5227 quantity article (09/07/2023)
Sea Urchin (*Paracentrotus lividus*)	Marine Biology, Herbicides, pescides and Ground Poisoning, Phylogenetics	2115 quantity article (09/07/2023
Ceriodaphnia dubia	Bioaccumulation, Microcystins, Nanoparticles, Stormwater, Macro invertabrates	1207 quantity article (09/07/2023)
Freshwater Ciliates	Marine Biology, Cell Biology, Parasitology, Bioengineering	1078 adet makale (09/07/2023)
Blowfish (*Fugu rubripes*)	Marine Biology, Immunology, Cancer and Development, Genetic, Chromosome disorders, Cell biology	731 quantity article (09/07/2023)
Chironomus sp. (*Mayflies*)	Marine Biology, Herbicides, pescides and Ground Poisoning, Molecular and Cell biology- Genetics	404 quantity article (09/07/2023)
Cyprinodon variegatus (*Sheepshead minnow*)	Herbicides, pescides and Ground Poisoning, Marine Biology, Phylogenetics and Genomics	310 quantity article (09/07/2023)
Macrostomum lignano	Animal sensing, Zoology and Animal ecology, Cell biology, Micro and Long Noncoding	146 quantity article (09/07/2023
Xenopus leavis	Cell biology-Physiology, Zoology and Animal Ecology, Sleep science, Circadian systems	79 quantity article (09/07/2023

3.2.2 Easy Cultivation

Model organisms can be easily cultivated because they have the potential for rapid reproduction and growth. This allows experiments to be carried out quickly.

3.2.3 Genetic and Biochemical Research

Model organisms can be integrated with tools and techniques used in genetic and biochemical studies. This makes it possible to perform more detailed analyses to understand the effects of chemicals on target molecules.

3.2.4 Lower Cost

The use of model organisms can reduce the cost of large-scale experiments. Instead of large and economically important species, smaller and easily available organisms can be used for more comprehensive studies.

However, the use of model organisms also has some limitations. Model organisms may not be fully representative of real-world conditions and results may not be directly applicable to other species. Therefore, it is important to use more than one species of organisms in aquatic toxicology research and to validate the results on different species.

3.3 Conclusion

Model organisms are an important research tool in aquatic toxicology studies. They are used to understand the effects of chemicals in the aquatic environment and provide important data for the protection of water resources. However, the applicability of results to real-world conditions should be considered and validation of results should be ensured by using a variety of organism species.

Acknowledgements Many thanks to my colleagues from Fisheries Faculty who supported me in the research for their help Atatürk University for their support in my academic studies.

References

Aalizadeh R, von der Ohe PC, Thomaidis NS (2017) Prediction of acute toxicity of emerging contaminants on the water flea Daphnia magna by Ant Colony Optimization–Support Vector Machine QSTR models. Environ Sci Proc Impacts 19(3):438–448. https://doi.org/10.1039/C6EM00679E

Abdel-Warith AA, Younis EM, Al-Asgah NA, Wahbi OM (2011) Effect of zinc toxicity on liver histology of Nile tilapia Oreochromis Niloticus. Scientific Res Essays 6(17):3760–3769. https://doi.org/10.5897/SRE11.883

Abraham JS, Sripoorna S, Choudhary A, Toteja R, Gupta R, Makhija S, Warren A (2017) Assessment of heavy metal toxicity in four species of freshwater ciliates (Spirotrichea: Ciliophora) from Delhi, India. Current Sci 2141–2150. https://doi.org/10.1007/s00203-017-1385-y

Alejo A, Tafalla C (2011) Chemokines in teleost fish species. Dev Comp Immunol 35(12):1215–1222. https://doi.org/10.1016/j.dci.2011.03.011

Anagha B, Athira PS, Anisha P, Charles PE, Anandkumar A, Rajaram R (2022) Biomonitoring of heavy metals accumulation in molluscs and echinoderms collected from southern coastal India. Mar Pollut Bull 184:114169. https://doi.org/10.1016/j.marpolbul.2022.114169

Andreassen R, Høyheim B (2017) MiRNAs associated with immune response in teleost fish. Dev Comp Immunol 75:77–85. https://doi.org/10.1016/j.dci.2017.02.023

Atamanalp M, Türkez H, Yeltekin AÇ, Özgeriş FB, Ucar A, Çağlar Ö, Alak G (2022) Borax relieved the acrylamide-induced hematotoxic, hepatotoxic, immunotoxic and genotoxic damages in rainbow trout by regulating apoptosis and Nrf2 signaling pathway. Comp Biochem Physiol C: Toxicol Pharmacol 259:109396. https://doi.org/10.1016/j.cbpc.2022.109396

Babayigit A, Duy Thanh D, Ethirajan A, Manca J, Muller M, Boyen HG, Conings B (2016) Assessing the toxicity of Pb-and Sn-based perovskite solar cells in model organism *Danio rerio*. Sci Rep 6(1):18721. https://doi.org/10.1038/srep18721

Bay S, Burgess R, Nacci D (1993) Status and applications of echinoid (Phylum Echinodermata) toxicity test methods. ASTM Spec Tech Publ 1179:281–281

Bergami E, Pugnalini S, Vannuccini ML, Manfra L, Faleri C, Savorelli F, Corsi IJAT (2017) Long-term toxicity of surface-charged polystyrene nanoplastics to marine planktonic species *Dunaliella tertiolecta* and *Artemia franciscana*. Aquat Toxicol 189:159–169. https://doi.org/10.1016/j.aquatox.2017.06.008

Bickmeyer U, Meinen I, Meyer S, Kröner S, Brenner M (2020) Fluorescence measurements of the marine flatworm Macrostomum lignano during exposure to TNT and its derivatives 2-ADNT and 4-ADNT. Mar Environ Res 161:105041. https://doi.org/10.1016/j.marenvres.2020.105041

Blouin MS, Wrey MC, Bollmann SR, Skaar JC, Twibell RG, Fuentes C (2021) Offspring of first-generation hatchery steelhead trout (*Oncorhynchus mykiss*) grow faster in the hatchery than offspring of wild fish, but survive worse in the wild: Possible mechanisms for inadvertent domestication and fitness loss in hatchery salmon. PLoS ONE 16(12):e0257407. https://doi.org/10.1371/journal.pone.0257407

Bodensteiner BL, Agudelo-Cantero GA, Arietta AA, Gunderson AR, Muñoz MM, Refsnider JM, Gangloff EJ (2021) Thermal adaptation revisited: How conserved are thermal traits of reptiles and amphibians? J Exp Zool Part a: Ecol Integr Phys 335(1):173–194. https://doi.org/10.1002/jez.2414

Boese BL, Lamberson JO, Swartz RC, Ozretich RJ (1997) Photoinduced toxicity of fluoranthene to seven marine benthic crustaceans. Arch Environ Contam Toxicol 32:389–393. https://doi.org/10.1007/s002449900201

Bonfim RC, Oliveira FAD, Godoy RLDO, Rosenthal A (2019) A review on high hydrostatic pressure for bivalve mollusk processing: relevant aspects concerning safety and quality. Food Sci Tech 39:515–523. https://doi.org/10.1590/fst.26918

Bonini NM, Berger SL (2017) The sustained impact of model organisms—in genetics and epigenetics. Genetics 205(1):1–4. https://doi.org/10.1534/genetics.116.187864

Bradford YM, Van Slyke CE, Ruzicka L, Singer A, Eagle A, Fashena D, Westerfield M (2022) Zebrafish information network, the knowledgebase for Danio rerio research. Genetics 220(4):iyac016. https://doi.org/10.1093/genetics/iyac016
Bruschetti M (2019) Role of reef-building, ecosystem engineering polychaetes in shallow water ecosystems. Diversity 11(9):168. https://doi.org/10.3390/d11090168
Buonanno F, Guella G, Strim C, Ortenzi C (2012) Chemical defence by mono-prenyl hydroquinone in a freshwater ciliate, *Spirostomum ambiguum*. Hydrobiologia 684:97–107. https://doi.org/10.1007/s10750-011-0972-1
Calmao M, Blasco N, Benito A, Thoppil R, Torre-Fernandez I, Castro K, Soto M (2023) Time-course distribution of fluorescent microplastics in target tissues of mussels and polychaetes. Chemosphere 311:137087. https://doi.org/10.1016/j.chemosphere.2022.137087
Canedo A, Saiki P, Santos AL, Carneiro KDS, Souza AMD, Qualhato G, Rocha TL (2022) Zebrafish (Danio rerio) meets bioethics: the 10Rs ethical principles in research. Ciência Animal Brasileira, 23:e-70884. https://doi.org/10.1590/1809-6891v22e-70884
Carey N, Chester ET, Robson BJ (2023) Loss of functionally important and regionally endemic species from streams forced into intermittency by global warming. Glob Change Biol 29(11):3019–3038. https://doi.org/10.1111/gcb.16650
Chen S, Jiang S, Jiang H (2020) A review on conversion of crayfish-shell derivatives to functional materials and their environmental applications. J Bioresources Bioproducts 5(4):238–247. https://doi.org/10.1016/j.jobab.2020.10.002
Clare EL, Barber BR, Sweeney BW, Hebert PDN, Fenton MB (2011) Eating local: influences of habitat on the diet of little brown bats (*Myotis lucifugus*). Mol Ecol 20(8):1772–1780. https://doi.org/10.1111/j.1365-294X.2011.05040.x
Connors KA, Brill JL, Norberg-King T, Barron MG, Carr G, Belanger SE (2022) *Daphnia magna* and *Ceriodaphnia dubia* have similar sensitivity in standard acute and chronic toxicity tests. Environ Toxicol Chem 41(1):134–147. https://doi.org/10.1002/etc.5249
da Silva AQ, de Souza Abessa DM (2019) Toxicity of three emerging contaminants to non-target marine organisms. Environ Sci Pollut Res 26:18354–18364. https://doi.org/10.1007/s11356-019-05151-9
Dahms HU, Hagiwara A, Lee JS (2011) Ecotoxicology, ecophysiology, and mechanistic studies with rotifers. Aquat Toxicol 101(1):1–12. https://doi.org/10.1016/j.aquatox.2010.09.006
D'Costa AH (2022) Microplastics in decapod crustaceans: Accumulation, toxicity and impacts, a review. Sci Total Environ 832:154963. https://doi.org/10.1016/j.scitotenv.2022.154963
do Amaral DF, Guerra V, Motta AGC, e Silva DDM, Rocha TL (2019) Ecotoxicity of nanomaterials in amphibians: a critical review. Sci Total Env 686:332–344. https://doi.org/10.1016/j.scitotenv.2019.05.487
dos Reis Oliveira PC, Kraak MH, Pena-Ortiz M, van der Geest HG, Verdonschot PF (2020) Responses of macroinvertebrate communities to land use specific sediment food and habitat characteristics in lowland streams. Sci Total Environ 703:135060. https://doi.org/10.1016/j.scitotenv.2019.135060
ElFessi R, Khamessi O, Srairi-Abid N, Sabatier JM, Tytgat J, Peigneur S, Kharrat R (2022) Purification and Characterization of Bot33: a non-toxic peptide from the venom of Buthus occitanus tunetanus Scorpion. Molecules 27(21):7278. https://doi.org/10.3390/molecules27217278
Eriksson-Kallio AM, Holopainen R, Koski P, Nousiainen A, Koskinen H, Kause A, Gadd T (2020) Susceptibility of rainbow trout to three different genogroups of infectious pancreatic necrosis virus. Dis Aquat Org 141:103–116. https://doi.org/10.3354/dao03512
Evenhuis JP, LaFrentz BR (2016) Virulence of Flavobacterium columnare genomovars in rainbow trout *Oncorhynchus mykiss*. Dis Aquat Org 120(3):217–224. https://doi.org/10.3354/dao03027
Fang J, Chen T, Pan Q, Wang Q (2018) Generation of albino medaka (Oryzias latipes) by CRISPR/Cas9. J Exp Zool B Mol Dev Evol 330(4):242–246. https://doi.org/10.1002/jez.b.22808
Firkins JL, Yu Z, Park T, Plank JE (2020) Extending Burk Dehority's perspectives on the role of ciliate protozoa in the rumen. Front Microbiol 11:123. https://doi.org/10.3389/fmicb.2020.00123

Forsgren E, Dupont S, Jutfelt F, Amundsen T (2013) Elevated CO_2 affects embryonic development and larval phototaxis in a temperate marine fish. Ecol Evol 3(11):3637–3646. https://doi.org/10.1002/ece3.709

Friedman M (2010) Explosive morphological diversification of spiny-finned teleost fishes in the aftermath of the end-Cretaceous extinction. Proc Roy Soc B: Biol Sci 277(1688):1675–1683. https://doi.org/10.1098/rspb.2009.2177

Gaillard S, Réveillon D, Mason PL, Ayache N, Sanderson M, Smith JL, Vogelbein WK (2023) Mortality and histopathology in sheepshead minnow (Cyprinodon variegatus) larvae exposed to pectenotoxin-2 and Dinophysis acuminata. Aquat Toxicol 257:106456. https://doi.org/10.1016/j.aquatox.2023.106456

García-Roger EM, Lubzens E, Fontaneto D, Serra M (2019) Facing adversity: dormant embryos in rotifers. Biol Bull 237(2):119–144. https://doi.org/10.1086/705701

Genevière AM, Derelle E, Escande ML, Grimsley N, Klopp C, Ménager C, Moreau H (2020) Responses to iron oxide and zinc oxide nanoparticles in echinoderm embryos and microalgae: uptake, growth, morphology, and transcriptomic analysis. Nanotoxicology 14(10):1342–1361. https://doi.org/10.1080/17435390.2020.1827074

Ghezelayagh A, Harrington RC, Burress ED, Campbell MA, Buckner JC, Chakrabarty P, Near TJ (2022) Prolonged morphological expansion of spiny-rayed fishes following the end-Cretaceous. Nature Ecology and Evolution 6(8):1211–1220. https://doi.org/10.1038/s41559-022-01801-3

Gilbert JJ (2022) Food niches of planktonic rotifers: Diversification and implications. Limnol Oceanogr 67(10):2218–2251. https://doi.org/10.1002/lno.12199

Gioacchini G, Maradonna F, Lombardo F, Bizzaro D, Olivotto I, Carnevali O (2010) Increase of fecundity by probiotic administration in zebrafish (*Danio rerio*). Reproduction (cambridge, England) 140(6):953–959. https://doi.org/10.1530/rep-10-0145

Gladyshev E, Meselson M (2008) Extreme resistance of bdelloid rotifers to ionizing radiation. Proc Natl Acad Sci 105(13):5139–5144. https://doi.org/10.1073/pnas.0800966105

Gomez-Chiarri M, Guo X, Tanguy A, He Y, Proestou D (2015) The use of-omic tools in the study of disease processes in marine bivalve mollusks. J Invertebr Pathol 131:137–154. https://doi.org/10.1016/j.jip.2015.05.007

Gulías P, Guerra-Varela J, Gonzalez-Aparicio M, Ricobaraza A, Vales A, Gonzalez-Aseguinolaza G, Sánchez L (2019) *Danio rerio* as model organism for adenoviral vector evaluation. Genes 10(12):1053. https://doi.org/10.3390/genes10121053

Hemmer MJ, Hemmer BL, Bowman CJ, Kroll KJ, Folmar LC, Marcovich D, Denslow ND (2001) Effects of p-nonylphenol, methoxychlor, and endosulfan on vitellogenin induction and expression in sheepshead minnow (*Cyprinodon variegatus*). Env Toxicol Chem: an Int J 20(2):336–343. https://doi.org/10.1002/etc.5620200214

Henry TB, Black MC (2007) Mixture and single-substance acute toxicity of selective serotonin reuptake inhibitors in Ceriodaphnia dubia. Environ Toxicol Chem: an Int J 26(8):1751–1755. https://doi.org/10.1897/06-265R.1

Huang AG, Tu X, Qi XZ, Ling F, Zhu B, Wang GX (2019) Gardenia jasminoides Ellis inhibit white spot syndrome virus replication in red swamp crayfish *Procambarus clarkii*. Aquaculture 504:239–247. https://doi.org/10.1016/j.aquaculture.2019.02.008

Ignoto S, Pecoraro R, Scalisi EM, Contino M, Ferruggia G, Indelicato S, Brundo MV (2023) Spermiotoxicity of Nano-TiO_2 compounds in the sea urchin *Paracentrotus lividus* (Lamarck, 1816): considerations on water remediation. J Mar Sci Eng 11(2):380. https://doi.org/10.3390/jmse11020380

Iigo M, Furukawa K, Nishi G, Tabata M, Aida K (2007) Ocular melatonin rhythms in teleost fish. Brain Behav Evol 69(2):114–121. https://doi.org/10.1159/000095200

Imangali N, Phan QT, Mahady G, Winkler C (2021) The dietary anthocyanin delphinidin prevents bone resorption by inhibiting Rankl-induced differentiation of osteoclasts in a medaka (*Oryzias latipes*) model of osteoporosis. J Fish Biol 98(4):1018–1030. https://doi.org/10.1111/jfb.14317

Jaja-Chimedza A, Gantar M, Gibbs PD, Schmale MC, Berry JP (2012) Polymethoxy-1-alkenes from Aphanizomenon ovalisporum inhibit vertebrate development in the zebrafish (Danio rerio) embryo model. Mar Drugs 10(10):2322–2336. https://doi.org/10.3390/md10102322

Jensen AA, Hammer H, Ebert B, Jensen HS (2012) Clobazam and ıts active metabolite, N-Desmethylclobazam, are partial benzodiazepine receptor agonists at cloned gabaa receptors expressed in Xenopus leavis Oocytes (P05. 085)

Jensen KM, Korte JJ, Kahl MD, Pasha MS, Ankley GT (2001) Aspects of basic reproductive biology and endocrinology in the fathead minnow (Pimephales promelas). Comp Biochem Physiol C: Toxicol Pharmacol 128(1):127–141. https://doi.org/10.1016/S1532-0456(00)00185-X

Jeppesen E, Leavitt P, De Meester L, Jensen JP (2001) Functional ecology and palaeolimnology: using cladoceran remains to reconstruct anthropogenic impact. Trends Ecol Evol 16(4):191–198. https://doi.org/10.1016/S0169-5347(01)02100-0

Jiang Y, Cao C (2021) Crayfish–rice integrated system of production: an agriculture success story in China. A Rev Agronomy Sustainable Dev 41:1–14. https://doi.org/10.1007/s13593-021-00724-w

Judith S, Candelmo A (2012) Pollutants and fish predator/prey behavior: a review of laboratory and field approaches. https://doi.org/10.1093/czoolo/58.1.9

Jung YJ, Kim KT, Kim JY, Yang SY, Lee BG, Kim SD (2014) Bioconcentration and distribution of silver nanoparticles in Japanese medaka (*Oryzias latipes*). J Hazard Mater 267:206–213. https://doi.org/10.1016/j.jhazmat.2013.12.061

Junker M, Rapoport TA (2015) Involvement of VAT-1 in phosphatidylserine transfer from the endoplasmic reticulum to mitochondria. Traffic 16(12):1306–1317. https://doi.org/10.1111/tra.12336

Kamijo M, Kojima K, Maruyama K, Konno N, Motohashi E, Ikegami T, Matsuda K (2011) Neuropeptide Y in tiger puffer (*Takifugu rubripes*): distribution, cloning, characterization, and mRNA expression responses to prandial condition. Zoolog Sci 28(12):882–890. https://doi.org/10.2108/zsj.28.882

Kawakami K, Shima A, Kawakami N (2000) Identification of a functional transposase of the Tol_2 element, an Ac-like element from the Japanese medaka fish, and its transposition in the zebrafish germ lineage. Proc Natl Acad Sci 97(21):11403–11408. https://doi.org/10.1073/pnas.97.21.11403

Kirby AR, Crossley DA, Mager EM (2020) The metabolism and swimming performance of sheepshead minnows (*Cyprinodon variegatus*) following thermal acclimation or acute thermal exposure. J Comp Physiol B 190:557–568. https://doi.org/10.1007/s00360-020-01293-2

Kuroda T, Ritchey CM, Podlesnik CA (2023) Selective effects of conspecific movement on social preference in zebrafish (*Danio rerio*) using real-time 3D tracking and 3D animation. Sci Rep 13(1):10502. https://doi.org/10.1038/s41598-023-37579-y

LaFrentz BR, LaPatra SE, Shoemaker CA, Klesius PH (2012) Reproducible challenge model to investigate the virulence of Flavobacterium columnare genomovars in rainbow trout *Oncorhynchus mykiss*. Dis Aquat Org 101(2):115–122. https://doi.org/10.3354/dao02522

Larson ER, Olden JD (2016) Field sampling techniques for crayfish. Biol Ecol Crayfish 287:324

Lee BY, Choi BS, Kim MS, Park JC, Jeong CB, Han J, Lee JS (2019) The genome of the freshwater water flea Daphnia magna: a potential use for freshwater molecular ecotoxicology. Aquat Toxicol 210:69–84. https://doi.org/10.1016/j.aquatox.2019.02.009

Leppänen JJ (2018) An overview of Cladoceran studies conducted in mine water impacted lakes. Int Aquat Res 10:207–221. https://doi.org/10.1007/s40071-018-0204-7

Lin Z, Hosoya S, Sato M, Mizuno N, Kobayashi Y, Itou T, Kikuchi K (2020) Genomic selection for heterobothriosis resistance concurrent with body size in the tiger pufferfish. Takifugu Rubripes Sci Rep 10(1):19976. https://doi.org/10.1038/s41598-020-77069-z

Lu K, Wu J, Tang S, Jia X, Liang XF (2023) Knockout of sws2a and sws2b in Medaka (Oryzias latipes) reveals their roles in regulating vision-guided behavior and eye development. Int J Mol Sci 24(10):8786. https://doi.org/10.3390/ijms24108786

Luo Y, Liu L, Yang P, Xia H, Zhao J, Chen N, Sun Y (2019) Advances in ıntestinal mucosal ımmunoglobulins of teleost fish. Israeli J Aquaculture-Bamidgeh, 71. http://hdl.handle.net/10524/62966

Ma Y, He J, Sieber M, von Frieling J, Bruchhaus I, Baines JF, Roeder T (2023) The microbiome of the marine flatworm Macrostomum lignano provides fitness advantages and exhibits circadian rhythmicity. Commun Biol 6(1):289. https://doi.org/10.1038/s42003-023-04671-y

Marshall WS, Lynch EM, Cozzi RRF (2002) Redistribution of immunofluorescence of CFTR anion channel and NKCC cotransporter in chloride cells during adaptation of the killifish Fundulus heteroclitus to sea water. J Exp Biol 205(9):1265–1273. https://doi.org/10.1242/jeb.205.9.1265

Matthiessen P (2008) An assessment of endocrine disruption in mollusks and the potential for developing internationally standardized mollusk life cycle test guidelines. Integr Environ Assess Manag 4(3):274–284. https://doi.org/10.1897/IEAM_2008-003.1

Metsoviti MN, Papapolymerou G, Karapanagiotidis IT, Katsoulas N (2019) Comparison of growth rate and nutrient content of five microalgae species cultivated in greenhouses. Plants 8(8):279. https://doi.org/10.3390/plants8080279

Miller W, Makova KD, Nekrutenko A, Hardison RC (2004) Comparative genomics. Annu Rev Genomics Hum Genet 5:15–56. https://doi.org/10.1146/annurev.genom.5.061903.180057

Minghetti M, Drieschner C, Bramaz N, Schug H, Schirmer K (2017) A fish intestinal epithelial barrier model established from the rainbow trout (Oncorhynchus mykiss) cell line, RTgutGC. Cell Biol Toxicol 33:539–555. https://doi.org/10.1007/s10565-017-9385-x

Moeller HV, Johnson MD (2023) Mesodinium. Curr Biol 33(7):R249–R250. https://doi.org/10.1016/j.cub.2023.02.015

Montagnes D, Roberts E, Lukeš J, Lowe C (2012) The rise of model protozoa. Trends Microbiol 20(4):184–191. https://doi.org/10.1016/j.tim.2012.01.007

Morroni L, Gaion A, Broccoli A, Ferrari S, Pellegrini D, Sartori D (2022) Influence of salinity on copper toxicity in paracentrotus lividus and arbacia lixula embryos. Water 15(1):65. https://doi.org/10.3390/w15010065

Mount DR, Erickson RJ, Forsman BB, Highland TL, Hockett JR, Hoff DJ, Norberg-King TJ (2019) Chronic toxicity of major ion salts and their mixtures to Ceriodaphnia dubia. Environ Toxicol Chem 38(4):769–783. https://doi.org/10.1002/etc.4346

Murashita K, Yoshiura Y, Chisada SI, Furuita H, Sugita T, Matsunari H, Yamamoto T (2014) Homologue gene of bile acid transporters ntcp, asbt, and ost-alpha in rainbow trout Oncorhynchus mykiss: tissue expression, effect of fasting, and response to bile acid administration. Fish Physiol Biochem 40:511–525. https://doi.org/10.1007/s10695-013-9862-y

Muths E, Hossack BR, Campbell Grant EH, Pilliod DS, Mosher BA (2020) Effects of snowpack, temperature, and disease on demography in a wild population of amphibians. Herpetologica 76(2):132–143. https://doi.org/10.1655/0018-0831-76.2.132

Naiel MA, Ismael NE, Abd El-hameed SA, Amer MS (2020) The antioxidative and immunity roles of chitosan nanoparticle and vitamin C-supplemented diets against imidacloprid toxicity on *Oreochromis niloticus*. Aquaculture 523:735219. https://doi.org/10.1016/j.aquaculture.2020.735219

Nassef M, Matsumoto S, Seki M, Khalil F, Kang IJ, Shimasaki Y, Honjo T (2010) Acute effects of triclosan, diclofenac and carbamazepine on feeding performance of Japanese medaka fish (Oryzias latipes). Chemosphere 80(9):1095–1100. https://doi.org/10.1016/j.chemosphere.2010.04.073

Nati JJ, Halsey LG, Johnson PC, Lindström J, Killen SS (2023) Does aerobic scope influence geographical distribution of teleost fishes? Conserv Physiol 11(1):coad012. https://doi.org/10.1093/conphys/coad012

Near TJ, Dornburg A, Eytan RI, Keck BP, Smith WL, Kuhn KL, Wainwright PC (2013) Phylogeny and tempo of diversification in the superradiation of spiny-rayed fishes. Proc Natl Acad Sci 110(31):12738–12743. https://doi.org/10.1073/pnas.1304661110

Newbold CJ, De La Fuente G, Belanche A, Ramos-Morales E, McEwan NR (2015) The role of ciliate protozoa in the rumen. Front Microbiol 6:1313. https://doi.org/10.3389/fmicb.2015.01313

Nipper MG, Prósperi VA, Zamboni AJ (1993) Toxicity testing with coastal species of southeastern Brazil. Echinoderm sperm and embryos. Bull Environ Contam Toxicol 50:646–652

Nobre CR, Santana MFM, Maluf A, Cortez FS, Cesar A, Pereira CDS, Turra A (2015) Assessment of microplastic toxicity to embryonic development of the sea urchin *Lytechinus variegatus* (Echinodermata: Echinoidea). Mar Pollut Bull 92(1–2):99–104. https://doi.org/10.1016/j.marpolbul.2014.12.050

Nys C, Janssen CR, Mager EM, Esbaugh AJ, Brix KV, Grosell M, De Schamphelaere KA (2014) Development and validation of a biotic ligand model for predicting chronic toxicity of lead to Ceriodaphnia dubia. Environ Toxicol Chem 33(2):394–403. https://doi.org/10.1002/etc.2433

Oliviero M, Tato T, Schiavo S, Fernández V, Manzo S, Beiras R (2019) Leachates of micronized plastic toys provoke embryotoxic effects upon sea urchin Paracentrotus lividus. Environ Pollut 247:706–715. https://doi.org/10.1016/j.envpol.2019.01.098

Orrego R, Guchardi J, Beyger L, Krause R, Holdway D (2011) Comparative embryotoxicity of pulp mill extracts in rainbow trout (*Oncorhynchus mykiss*), American flagfish (Jordanella floridae) and Japanese medaka (*Oryzias latipes*). Aquat Toxicol 104(3–4):299–307. https://doi.org/10.1016/j.aquatox.2011.04.015

Orsini L, Spanier KI, De Meester LUC (2012) Genomic signature of natural and anthropogenic stress in wild populations of the waterflea Daphnia magna: validation in space, time and experimental evolution. Mol Ecol 21(9):2160–2175. https://doi.org/10.1111/j.1365-294X.2011.05429.x

Park JC, Hagiwara A, Park HG, Lee JS (2020) The glutathione S-transferase genes in marine rotifers and copepods: Identification of GSTs and applications for ecotoxicological studies. Mar Pollut Bull 156:111080. https://doi.org/10.1016/j.marpolbul.2020.111080

Parlak V (2018) Evaluation of apoptosis, oxidative stress responses, AChE activity and body malformations in zebrafish (*Danio rerio*) embryos exposed to deltamethrin. Chemosphere 207:397–403. https://doi.org/10.1016/j.chemosphere.2018.05.112

Parrella A, Lavorgna M, Criscuolo E, Russo C, Fiumano V, Isidori M (2014) Acute and chronic toxicity of six anticancer drugs on rotifers and crustaceans. Chemosphere 115:59–66. https://doi.org/10.1016/j.chemosphere.2014.01.013

Patrikainen MS, Tolvanen ME, Aspatwar A, Barker HR, Ortutay C, Jänis J, Parkkila S (2017) Identification and characterization of a novel zebrafish (*Danio rerio*) pentraxin–carbonic anhydrase. PeerJ 5:e4128. https://doi.org/10.7717/peerj.4128

Peterson TS, Ferguson JA, Watral VG, Mutoji KN, Ennis DG, Kent M (2013) Paramecium caudatum enhances transmission and infectivity of *Mycobacterium marinum* and *M. chelonae* in zebrafish *Danio rerio*. Diseases of aquatic organisms, 106(3):229–239. https://doi.org/10.3354/dao02649

Prabhakar AK, Mohan BC, Tay TS, Lee SSC, Teo SLM, Wang CH (2021) Incinerated sewage sludge bottom ash-chemical processing, leaching patterns and toxicity testing. J Hazard Mater 402:123350. https://doi.org/10.1016/j.jhazmat.2020.123350

Raimondo S, Hemmer BL, Goodman LR, Cripe GM (2009) Multigenerational exposure of the estuarine sheepshead minnow (Cyprinodon variegatus) to 17β-Estradiol. II. Population-level effects through two life cycles. Environ Toxicol Chem: An Int J 28(11):2409–2415. https://doi.org/10.1897/08-540.1

Raimondo S, Hemmer BL, Lilavois CR, Krzykwa J, Almario A, Awkerman JA, Barron MG (2016) Effects of Louisiana crude oil on the sheepshead minnow (*Cyprinodon variegatus)* during a life-cycle exposure to laboratory oiled sediment. Environ Toxicol 31(11):1627–1639. https://doi.org/10.1002/tox.22167

Rainbow PS, Luoma SN (2011) Metal toxicity, uptake and bioaccumulation in aquatic invertebrates—modelling zinc in crustaceans. Aquat Toxicol 105(3–4):455–465. https://doi.org/10.1016/j.aquatox.2011.08.001

Raza T, Rasool B, Asrar M, Manzoor M, Javed Z, Jabeen F, Younis T (2023) Exploration of polyacrylamide microplastics and evaluation of their toxicity on multiple parameters of *Oreochromis niloticus*. Saudi J Biol Sci 30(2):103518. https://doi.org/10.1016/j.sjbs.2022.103518

Romano N, Zeng C (2013) Toxic effects of ammonia, nitrite, and nitrate to decapod crustaceans: a review on factors influencing their toxicity, physiological consequences, and coping mechanisms. Rev Fish Sci 21(1):1–21. https://doi.org/10.1080/10641262.2012.753404

Salvo A, Cicero N, Vadalà R, Mottese AF, Bua D, Mallamace D, Dugo G (2016) Toxic and essential metals determination in commercial seafood: *Paracentrotus lividus* by ICP-MS. Nat Prod Res 30(6):657–664. https://doi.org/10.1080/14786419.2015.1038261

Santos-Ballardo DU, Rossi S, Hernández V, Gómez RV, del Carmen R-U, Caro-Corrales J, Valdez-Ortiz A (2015) A simple spectrophotometric method for biomass measurement of important microalgae species in aquaculture. Aquaculture 448:87–92. https://doi.org/10.1016/j.aquaculture.2015.05.044

Schirmer E, Ritschar S, Ochs M, Laforsch C, Schuster S, Römpp A (2022) MALDI mass spectrometry imaging workflow for the aquatic model organisms *Danio rerio* and *Daphnia magna*. Sci Rep 12(1):7288. https://doi.org/10.1038/s41598-022-09659-y

Shima A, Mitani H (2004) Medaka as a research organism: past, present and future. Mech Dev 121(7–8):599–604. https://doi.org/10.1016/j.mod.2004.03.011

Shomorin GO, Storebakken T, Kraugerud OF, Øverland M, Hansen BR, Hansen JØ (2019) Evaluation of wedge wire screen as a new tool for faeces collection in digestibility assessment in fish: The impact of nutrient leaching on apparent digestibility of nitrogen, carbon and sulphur from fishmeal, soybean meal and rapeseed meal-based diets in rainbow trout (*Oncorhynchus mykiss*). Aquaculture 504:81–87. https://doi.org/10.1016/j.aquaculture.2019.01.051

Šiller L, Lemloh ML, Piticharoenphun S, Mendis BG, Horrocks BR, Brümmer F, Medaković D (2013) Silver nanoparticle toxicity in sea urchin *Paracentrotus lividus*. Environ Pollut 178:498–502. https://doi.org/10.1016/j.envpol.2013.03.010

Smith AD, Bharath A, Mallard C, Orr D, Smith K, Sutton JA, Ozburn GW (1991) The acute and chronic toxicity of ten chlorinated organic compounds to the American flagfish (Jordanella floridae). Arch Environ Contam Toxicol 20:94–102. https://doi.org/10.1007/BF01065334

Snell TW, Janssen CR (1995) Rotifers in ecotoxicology: a review. Hydrobiologia 313:231–247. https://doi.org/10.1007/BF00025956

Sommer U, Sommer F (2006) Cladocerans versus copepods: the cause of contrasting top–down controls on freshwater and marine phytoplankton. Oecologia 147(2):183–194. https://doi.org/10.1007/s00442-005-0320-0

Stachurska A, Sarna T (2019) Regulation of melanopsin signaling: key interactions of the nonvisual photopigment. Photochem Photobiol 95(1):83–94. https://doi.org/10.1111/php.12995

Stosik M, Tokarz-Deptuła B, Deptuła W (2021) Immunological memory in teleost fish. Fish Shellfish Immunol 115:95–103. https://doi.org/10.1016/j.fsi.2021.05.022

Tietje K, Rivera-Ingraham G, Petters C, Abele D, Dringen R, Bickmeyer U (2013) Reporter dyes demonstrate functional expression of multidrug resistance proteins in the marine flatworm Macrostomum lignano: the sponge-derived dye Ageladine A is not a substrate of these transporters. Mar Drugs 11(10):3951–3969. https://doi.org/10.3390/md11103951

Ucar A, Arslan ME, Cilingir Yeltekin A, Ozgeris FB, Caglar Yıldırım O, Parlak V, Atamanalp, M (2023) Neutralization of iron oxide magnetic nanoparticle aquatoxicity on *Oncorhynchus mykiss* via supplementation with ulexite. Drug and Chem Toxicol 1–13.https://doi.org/10.1080/01480545.2022.2164298

Uemura N, Koike M, Ansai S, Kinoshita M, Ishikawa-Fujiwara T, Matsui H, Takahashi R (2015) Viable neuronopathic Gaucher disease model in Medaka (*Oryzias latipes*) displays axonal accumulation of alpha-synuclein. PLoS Genet 11(4):e1005065. https://doi.org/10.1371/journal.pgen.1005065

Valenti TW Jr, Perez-Hurtado P, Chambliss CK, Brooks BW (2009) Aquatic toxicity of sertraline to Pimephales promelas at environmentally relevant surface water pH. Environ Toxicol Chem 28(12):2685–2694. https://doi.org/10.1897/08-546.1

Verdes A, Anand P, Gorson J, Jannetti S, Kelly P, Leffler A, Holford M (2016) From mollusks to medicine: a venomics approach for the discovery and characterization of therapeutics from Terebridae peptide toxins. Toxins 8(4):117. https://doi.org/10.3390/toxins8040117

Vilas-Boas JA, Cardoso SJ, Senra MVX, Rico A, Dias RJP (2020) Ciliates as model organisms for the ecotoxicological risk assessment of heavy metals: a meta-analysis. Ecotoxicol Environ Saf 199:110669. https://doi.org/10.1016/j.ecoenv.2020.110669

Wallace RL (2002) Rotifers: exquisite metazoans. Integr Comp Biol 42(3):660–667. https://doi.org/10.1093/icb/42.3.660

Wang WX, Lu G (2017) Heavy metals in bivalve mollusks. Chemical contaminants and residues in food, pp 553–594. https://doi.org/10.1016/B978-0-08-100674-0.00021-7

Wang XA, Ma A, Zhao Y, Yue L, Sun J, Meng X, Liu S (2015) Prediction of Takifugu rubripes weight based on radial basis function neural network. J Fisher China 39(12):1799–1806. http://www.scxuebao.cn

Weidner S, Latz E, Agaras B, Valverde C, Jousset A (2017) Protozoa stimulate the plant beneficial activity of rhizospheric pseudomonads. Plant Soil 410:509–515. https://doi.org/10.1007/s11104-016-3094-8

Winder VL, Pennington PL, Hurd MW, Wirth EF (2012) Fluoxetine effects on sheepshead minnow (Cyprinodon variegatus) locomotor activity. J Environ Sci Health B 47(1):51–58. https://doi.org/10.1080/03601234.2012.607767

Wolters JW, Verdonschot RC, Schoelynck J, Brion N, Verdonschot PF, Meire P (2018) Stable isotope measurements confirm consumption of submerged macrophytes by macroinvertebrate and fish taxa. Aquat Ecol 52:269–280. https://doi.org/10.1007/s10452-018-9662-7

Xiong JQ, Kurade MB, Jeon BH (2017) Ecotoxicological effects of enrofloxacin and its removal by monoculture of microalgal species and their consortium. Environ Pollut 226:486–493. https://doi.org/10.1016/j.envpol.2017.04.044

Yanbo W, Wenju Z, Weifen L, Zirong X (2006) Acute toxicity of nitrite on tilapia (Oreochromis niloticus) at different external chloride concentrations. Fish Physiol Biochem 32:49–54. https://doi.org/10.1007/s10695-005-5744-2

Zadesenets KS, Schärer L, Rubtsov NB (2017) New insights into the karyotype evolution of the free-living flatworm Macrostomum lignano (Platyhelminthes, Turbellaria). Sci Rep 7(1):6066. https://doi.org/10.1038/s41598-017-06498-0

Zhang HQ, Jin XY, Li XP, Li MF (2023) IL8 of Takifugu rubripes is a chemokine that interacts with peripheral blood leukocytes and promotes antibacterial defense. Fish and Shellfish Immunology 108918. https://doi.org/10.1016/j.fsi.2023.108918

Chapter 4
Freshwater Toxicity Tests and Experimental Environment Procedures

Cristina Maria Ponepal, Nicoleta Anca Şuţan, Daniela Bărbuceanu, Alina Păunescu, Diana Ionela Stegăruş, and Liliana Cristina Soare

Abstract Contamination of aquatic ecosystems, including freshwaters, with varied and new polluting agents is a constant threat and a primary concern worldwide. Due to domestic, industrial or agricultural activities, a multitude of chemical pollutants can be found in freshwater, such as pharmaceuticals and personal care products, nanomaterials, plasticizers, endocrine disruptors, flame retardants, pesticides, persistent organic pollutants, etc. They affect aquatic fauna and flora with the most serious harmful consequences for food chains. Having a key role in ecosystems, plants from different taxonomic groups (algae, bryophytes, pteridophytes, flowering plants) are important test organisms for evaluating water quality and for its decontamination. Invertebrates, fish and amphibians through the distinct sensitivity of various life stages, the available and abundance, and the wide range of biomarkers that can be addressed in freshwater toxicity assessment have allowed for the design and standardization of a wide range of tests presented in this chapter, with an emphasis on the tests

C. M. Ponepal · N. A. Şuţan (✉) · D. Bărbuceanu · A. Păunescu · L. C. Soare
Department of Natural Sciences, National University of Science and Technology POLITEHNICA Bucharest, Piteşti University Center, 1 Targu din Vale Str., 110040 Pitesti, Romania
e-mail: nicoleta_anca.sutan@upb.ro

C. M. Ponepal
e-mail: maria.ponepal@upb.ro

D. Bărbuceanu
e-mail: daniela.barbuceanu@upb.ro

A. Păunescu
e-mail: alina.paunescu@upb.ro

L. C. Soare
e-mail: liliana.soare@upb.ro

D. I. Stegăruş
National Research and Development Institute for Cryogenics and Isotopic Technologies—ICSI Ramnicu Valcea, 4th Uzinei Street, 240050 Ramnicu Valcea, Romania
e-mail: diana.stegarus@icsi.ro

M. Atamanalp et al., *Aquatic Toxicology in Freshwater*, Springer Water,
https://doi.org/10.1007/978-3-031-56669-1_4

developed by the OECD and the US EPA. A brief reference is assigned to ecotoxicological modeling, cytotoxicity and genotoxicity assays. Despite the consistent number of toxicity tests intended to evaluate specific adverse outcomes, the conditions that must be met so that the results are reliable and ensure the extrapolation of the data are of great interest for tackling the major threats to freshwater ecosystems.

Keywords Ecotoxicological tests · Plant · Invertebrates · Fish · Tadpoles

List of Abbreviations

AEC	Actual environmental concentration
AFNOR	Association Française de normalisation
AMA	Amphibian metamorphosis assay
ASTM	American society for testing and materials
BAF	Bioaccumulation factor
BCF	Bioconcentration factor
BSAF	Biota-sediment accumulation factor
CeO_2NPs	Cerium oxide nanoparticles
CYP1A	Cytochrome
EC50	Average effective concentration
ECETOC	European center for ecotoxicology and toxicology of chemicals
EEC	Estimated environmental concentrations
EU	European union
FETAX	Frog embryo teratogenesis assay—xenopus
HPT	Hypothalamic-pituitary-thyroid
IC50	Average inhibition concentration
ISO	The international organization for standardization
KOW	Octanol-water partition coefficient
LAGDA	Amphibian larval development test
LC50	Average lethal concentration
LD50	Average lethal dose
LOEC	Lowest observed effect concentration
LOEL/LOAEL	Lowest observed (adverse) effect level
MAC	Maximum acceptable concentration
MT	Metallothionein
NOEC	No observed effect concentration
NOEL/NOAEL	No observed (adverse) effect level
NPDES	National pollutant discharge elimination system
OECD	Organization for economic cooperation and development
PCB	Polychlorinated diphenyl's
QSAR	Quantitative structure activity relationship
REACH	Registration, evaluation, authorization, and restriction of chemicals

SARs	Structure activity relationships
SQT	Sediment quality triad
SSDs	Species sensitivity distributions
TL50	Lethal mean time
TU	Toxicity units
US EPA	U.S. environmental protection agency
VTG	Vitellogenin
WET	Whole effluent tests
WFD	Water framework directive (European Commission 2000)
WT	Water tests

4.1 Introduction

Aquatic toxicology or aquatic toxicity testing is defined as the qualitative and quantitative study of toxic effects of environmental pollutants on aquatic systems and organisms (Adams 1995).

Several agencies and international organizations such as the U.S. Environmental Protection Agency (EPA), the European Union (EU), Water Framework Directive (WFD, European Commission 2000) and Registration, Evaluation, Authorization, and Restriction of Chemicals (REACH) require test data on ecotoxicological chemical effects (Lillicrap et al. 2016; Mitchell et al. 2002; Walker 2008). Before approvement for large-scale manufacturing and utilization, any chemical substance, mixture or newly produced material, is mandatorily tested in terms of the effects on the environment (Walker 2008).

Estimating the ecological impact of chemicals on freshwater ecosystems is a difficult task. There are still many uncertainties about the most appropriate test procedures for environmental risk assessment (Rand 2008). It is a sine qua non that without abiotic and/or biotic risk assessment the health of freshwater ecosystems cannot be predicted and maintained (Poikane et al. 2020).

Toxicological research methods assess the toxicity of a substance or a mixture according to the intensity of their action on aquatic organisms. These include chemical methods, ecological methods and ecotoxicological methods (Gavrilescu 2008).

The techniques for the chemical analysis of water have become more and more precise and sensitive (gas chromatography with mass detector for organic micropollutants; plasma mass spectrometry for heavy metals) allowing the detection of very small amounts of pollutants present in the aquatic environment (Alkas et al. 2017; Gavrilescu 2008; Skou et al. 2018).

Ecological methods consist in assessing the harmful action of some wastewater on the biocenoses in the receiving basin and study solely the final result of the selective action exerted by the toxic substances/polluting factors on the populations in the ecosystem (Gavrilescu 2008). Among the polluting factors we mention organic solvents, heavy metals, potential carcinogens, mutagens, pesticides, antimicrobials, metabolic modulators, all of which affect aquatic ecosystems through direct discharge, entrainment in water flow or groundwater contamination; insecticides, for example, are prone to leach in a proportion of 20% (Miyoshi et al. 2003; Sánchez-Bayo et al. 2013).

Biotoxicology assesses the interaction of toxic substances with sensitive biological systems, i.e., absorption and assimilation, distribution, metabolism and excretion, and further with macro-biological systems, i.e., bioaccumulation, biomagnification, bioconcentration (Lyubenova and Boteva 2016a, b). Environmental contaminants are able to modify the structure, productivity and even the trophic network of the biocenoses (Oroş 2002).

Ecotoxicological tests ensure the qualitative and quantitative estimation of the specific plants/animals/microorganism's reaction level determined by the disruptive action of some chemicals with toxic potential (Petre and Teodorescu 2009). Ecotoxicological studies, from biotests with invertebrates, algae and bacteria to multispecies tests, in situ tests and modelling tests, allow the predictability of the effects of pollutants in acute and chronic tests, in the short- and long-term exposure, respectively, to the potential toxicants, and on various levels of biological organization (Anderson et al. 2004; Petre and Teodorescu 2009; Rand 2008). The ecotoxicological tests must take into account the following three types of potential hazards that will have to be determined: toxicity, biodegradability and, respectively, the capacity for bioaccumulation/bioconcentration.

Toxicity indicators used in toxicological tests reveal the degree of chemical threat towards all levels of organization of living matter. Therefore, toxicity tests have four major objectives:

- to point out the toxicity and route of exposure to various chemical substances;
- to estimate and enable the management of risks generated by different chemical substances;
- to indicate environmental standards and regulations;
- to rank chemicals according to their toxicity to various species (Rand 2008).

Nevertheless, most ecotoxicological tests only provide an assessment of the dose or concentration that elicits a 50% specific response or the highest concentration of a chemical that produce no effect (Petre and Teodorescu 2009).

4.2 Toxicological Indicators

Toxicity tests are fundamental in assessing the ecological integrity of freshwater systems. Assessing the impact of chemicals or complex chemical mixtures is of high-priority in determining potential adverse effects on aquatic organisms and further through the transfer to other links of the food chains (Nair and Perumal 2002; Anderson et al. 2004). Two categories of toxicological indicators are used to measure toxicity.

4.2.1 Indicators of Acute Toxicity

In acute aquatic toxicity assessment, the basic indicator is the amount of toxic substance that induces a biological response in 50% of the test subject population:

- average lethal dose (LD50) expresses the amount of toxic substance that causes death in 50% of test subjects (mg/kg);
- average lethal concentration (LC50) is the best known and most used quantitative parameter of acute toxicity and represents the concentration of test chemicals (mg/L or μg/L) that kill 50% of an exposed population;
- lethal mean time (TL50) expresses the exposure time (hours or minutes) during which 50% of the subjects treated with a certain concentration or dose of the toxic substance die;

Considering that the effect may be other than mortality, such as immobilizing *Daphnia* sp. or the diminution of algae multiplication, the indicator could be:

- average effective concentration (EC50) which represent the amount of toxic substance that produces the biological response in 50% of test subjects;
- average inhibition concentration (IC50) which indicate the amount of toxic substance that produces a 50% inhibition of the intended effect.

Acute toxicity can be evaluated by tests performed on unicellular organisms (bacteria, algae, protozoa), cell cultures, as well as on various plant or animal species, such as algae, worms, insects, crustaceans, fish, mammals (Adams and Rowland 2002; Gavrilescu 2008).

4.2.2 Indicators of Chronic Toxicity

The EC_{50} is a generic indicator, which is also determined in chronic toxicity tests with the specification of effects or consequences. For example, EC_{50} is used to measure the toxic concentration that inhibits the multiplication of algae by 50%, that reduces daphnia offspring by 50%, that diminish by 50% the growth of biomass in lentils or reduces by 50% the peroxidase activity.

In order to establish the accepted limits of toxicity, the following indicators are used as regulatory instruments:

- No Observed Effect Concentration (NOEC) is the highest tested concentration at which no observable effect occurs. It is also used under the name No Observed Adverse Effect Concentration (NOAEC) to specify that it is about negative effects;
- Lowest Observed Effect Concentration (LOEC), also used under the name Lowest Observed Adverse Effect Concentration (LOAEC) is the lowest tested concentration at which an observable effect is produced;
- No Observed (Adverse) Effect Level (NOEL/NOAEL) is the highest level or dose at which no harmful effect is observed;

- Lowest Observed (Adverse) Effect Level (LOEL/LOAEL) is the highest level or dose at which a harmful effect is observed.

These indicators are commonly used in environmental protection in order to establish a regulatory indicator called the maximum acceptable concentration (MAC). The MAC is calculated as the geometric mean of the NOEC and LOEC values to which a safety factor of 10, 100 or 1000 is added. The safety factor expresses the differences between the tested organisms and the higher organisms (humans, animals, plants) to which the estimation is performed. In addition, the safety factor must indicate the sensitivity of various species, the heterogeneity of each population, the synergistic effect of other pollutants in the environment and the environmental changes following the bioavailability adjustment (Gavrilescu 2008; Tamba-Berehoiu 2014).

4.3 Criteria for Designing and Performing an Aquatic Toxicity Test

The design of a toxicological study should be based on its objectives, on the physicochemical properties and mode of action of the targeted substance, on the pattern of use, the state of the environment, on the test organism and its phenotype (Gavrilescu 2008; Girling et al. 2000).

The chemical's structural formula, purity, water and light stability, vapor pressure, biodegradability, etc. determine the distribution and kinetics in the environment of a chemical substance (Adams and Rowland 2002). Due to a particular mode of action, most chemicals that occur in freshwater have a specific target and induce certain downstream effects. Knowing the mode of action of a chemical provides the possibility to identify the most important effects to be measured during and after a toxicity test.

The test organisms must be healthy and originate from a homogeneous population (age, life stage, weight, size, etc.), unless otherwise indicated in the test instructions. The number of test organisms per concentration of tested substance and for the control lot must not be less than three; the sample size depends on the type and objective of the study. Species that can co-exist in a body of water represent the test organisms when the evaluation of the effect of a chemical substance on a community of aquatic species is intended. A negative control performed under the same conditions as the treatment group and lacking the chemical to be tested, is mandatory for all toxicity tests. The use of a specific solvent requires its addition to the control (Adams and Rowland 2002).

The experimental conditions (temperature, light, oxygen concentration, water hardness) and the concentration of the substance under evaluation must be measured and accurately maintained throughout the study to avoid their impact on its toxicity and/or availability.

The criteria used to design an aquatic toxicity test are (Adams and Rowland 2002; Girling et al. 2000; Mitchell et al. 2002; Rand 2008):

- to be broadly welcomed by the scientific community;
- to be standardized to ensure the replicability of the results;
- to be accessible and cost-effective;
- to use well-known model organism;
- to cover a range of concentrations, some of which are found in the environment;
- to include a realistic and manageable exposure period;
- to provide results that allow statistical interpretation;
- to provide data that can be used for risk estimation;
- to be sensitive enough to disclose and measure the investigated toxic effects;
- to predict the potential effects on similar species.

Carrying out a toxicity test involves identifying one or more available Estimated Environmental Concentrations (EEC) or Actual Environmental Concentration (AEC) of a chemical. EECs are consistently acquired through computer modelling or simple predictions (Rand 2008).

The major standardized aquatic tests were developed by the Organization for Economic Cooperation and Development (OECD) and EPA. Several aquatic test protocols were also released by The International Organization for Standardization (ISO) and the American Society for Testing and Materials (ASTM).

However, it is obvious that not all tests fulfill all the above criteria, due to the features that define the various toxicity tests. Good design and organization of the experiment assumes that the results obtained are valid, relevant and reproducible. Reproducibility is ensured by carrying out experiments in similar laboratory conditions in triplicate, microcosm tests in duplicate, without express mentions regarding complex macrocosm tests.

The actual tests are preceded by the so-called "limit tests" in which a number of organisms (the same number for each sample) are exposed to a single and appropriate "limit concentration", which for most industrial chemicals is represented by 100 mg/L or the limits of water solubility or dispersion.

Toxicological tests carried out under laboratory conditions present a number of shortcomings (Mitchell et al. 2002):

- organisms in nature are simultaneously exposed to several chemical substances and stress factors with potential synergistic or antagonistic effects, thus generating unexpected results;
- most laboratory toxicity tests are performed with a single chemical and under favorable conditions;
- in nature, organisms are not always continuously exposed to chemical substances; long-term studies conducted in the laboratory ensure continuous exposure throughout the life cycle of the test organism to the evaluated substance;
- the concentration and availability of the chemical may vary in the environment while remaining constant in the laboratory;
- most toxicological tests involve dissolving/adding the chemical into water; in nature, organisms are exposed to chemicals through food and sediment as well;
- in situ the biotope and biocenosis can guide the toxic effects; most laboratory tests are performed on a single species, ignoring biotope-biocenosis interactions;

- studies are carried out in the laboratory on healthy and similar organisms; in nature, organisms are defined by genetic variability and are physiologically more diverse.

4.4 Test Organisms in Freshwater Ecotoxicology

A wide variety of organisms (algae, crustaceans, fish, oligochaetes, daphnia and other small invertebrate aquatic organisms), representing different trophic levels, are used in acute and chronic toxicity studies (Lyubenova and Boteva 2016a, b).

In order to be used in toxicological tests, a species must meet several criteria (Gavrilescu 2008; Petre and Teodorescu 2009):

- have high sensitivity to different chemical substances;
- be readily available and abundant;
- be easy to grow and maintain in culture in the laboratory;
- have a high survival rate under the set up and controlled conditions of laboratory;
- be well known from the point of view of its biology, physiology, genetics and behavior.

Hypothetically, any aquatic organism can be used in aquatic ecotoxicology tests. In practice however, depending on the type of toxic substance and the affected ecosystem, one or more species in different life stages and with different sensitivity are selected, without there being a universally valid species and regardless of the toxicity test (Rand 2008).

Based on the specialized toxicology works published in the period 2010–2016, Lyubenova and Boteva (2016a, b) reported that common test organisms are crustaceans and fish (22% and 20%, respectively), insects (9%), mollusks (9%) and algae (8%). The most used freshwater species used in ecotoxicological studies were *Pseudokirchneriella subcapitata, Lemna spp., Daphnia magna, Dreissena polymorpha, Mytilus edulis, Danio rerio* and *Cyprinus carpio.*

4.5 Classification of Toxicological Tests

A wide spectrum of toxicity testing procedures is currently available. Toxicity tests can be classified according to several criteria such as the species exposed to test material, medium of exposure, cost, precision, complexity, the skill needed to conduct them and the effects they measure, as described below and summarized in Fig. 4.1.

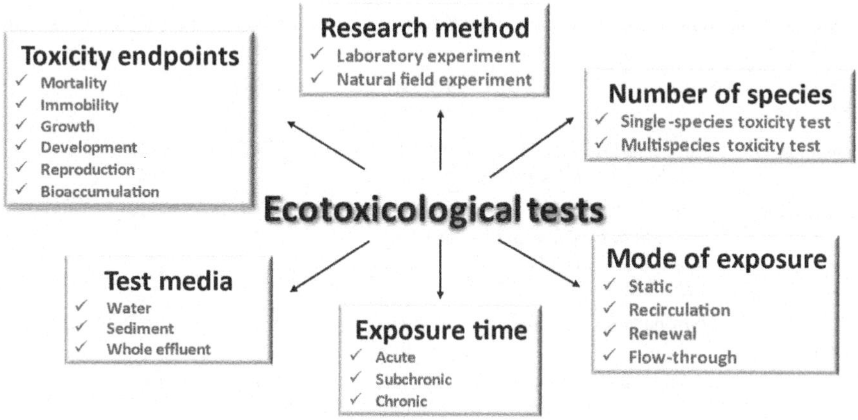

Fig. 4.1 Categories of aquatic toxicity tests

4.5.1 Classification of Toxicological Tests Based on the Recorded Effects

Depending on the test organism, each toxicity test is defined by one or more endpoints such as: survival, growth and development, reproduction capacity, immobilization, respiration, endocrine effects, and chemical bioaccumulation. The main toxicity endpoints usually determined in toxicity tests are shown in Table 4.1.

Bioaccumulation is often addressed in risk assessment and involves the determination of the octanol–water partition coefficient (KOW), bioconcentration factor (BCF), bioaccumulation factor (BAF) and biota-sediment accumulation factor (BSAF). In bioconcentration tests, the test organisms are exposed to a certain substance in a concentration relevant to the environment (less than one tenth of the species—specific LC50), and samples are analyzed periodically, until the steady state is reached (tissue residue levels remain constant) or for 28 days (Arnot and Gobas 2004; Jamil Emon et al. 2023; OECD Test No. 305 2012b, Rand 2008). After exposure the individuals are placed in clean water and the tissues depuration is assessed at various time intervals (the steady state is indicated by similar concentrations recorded in successive measurements made over 2 to 4 days).

Bioaccumulation results have been reported for different tissues and organs, such as the liver, kidney, gills, embryo tissue and accessory glands (Ali and Khan 2018; Jamil Emon et al. 2023; Li et al. 2015; Streit 1998). Bioaccumulation tests use BCF (the ratio of the average concentration of test chemical accumulated in the tissues of the test organism, under steady state conditions, to the average measured concentration in the water) to predict concentrations of hydrophobic contaminants in organisms.

The physiological state, the age and body size of the individuals of a bioindicator population determine the rates of accumulation of various chemical substances. In *Gammarus pulex* polychlorinated diphenyl's (PCB) content varied according to the

Table 4.1 The main toxicity endpoints for different aquatic organisms following exposure to chemicals

Organisms' category	Species	Toxicity endpoints	References
Cianobacteria	*Anabaena flos-aquae, Synechococcus leopoliensis*	Inhibition of growth	OECD 201 (2011a)
Microalgae	*Pseudokirchneriella subcapitata, Desmodesmus subspicatus, Navicula pelliculosa*	Inhibition of growth	OECD 201 (2011a)
Vascular plants	*Myriophyllum spicatum*	Vegetative growth	OECD 238, OECD 239 (2014a)
	Lemna gibba, L. minor	Vegetative growth	OECD 221 (2006)
Annelids	*Lumbriculus variegatus*	Growth, fecundity, bioaccumulation	OECD 225 (2007)
Mollusks	*Lymnaea stagnalis, Potamopyrgus antipodarum*	Mortality, growth, fecundity, bioconcentration	OECD 242 (2016a), OECD 243 (2016b)
Arthropods	*Daphnia magna, Daphnia pulex* *Ceriodaphnia dubia, Hyalella azteca, Gammarus fasciatus, Gammarus pseudolimnaeus, Gammarus lacustris, Chironomus riparuis, Chironomus tentans. Chironomus dilutus*	Mortality, growth, immobility, fecundity, development	OECD 202 (2004), OECD 218 (2018), OECD 219 (2018), OECD 233 (2010), OECD 235 (2011c), ISO 634 (2012), OECD 211 (2012c), EPA-OCSPP 850.1300 (2016a; b), EPA 1002.0 2002a, b, c, d, e, EPA-HQ-OPPT-2009-0154-0040 (2015), OCSPP 850.1020 (2016a)
Fish	*Pimephales promelas* *Brachidanio rerio* *Oryzias latipes* *Oncorhynchus mykiss* *Cyprinus carpio* *Carassius auratus* *Lepomis macrochirus*	Mortality, loss of swimming equilibrium, developmental deformities and skeletal abnormalities, length, yolk coagulation, growth, tumors, reproductive maturation, fecundity, histopathology, egg hatching, behavior, etc.	OECD 203 (2019), OECD 210 (2013a), OECD 212 (1998), OECD 215 (2000), OECD 229 (2012a), OECD 230 (2009a), OECD 234 (2011b), OECD 236 (2013b), OECD 240 (2023), OECD 305 (2012b), EPA 1001.0 (2002), EPA 1000.0 (2002), MEOGRT (2015), EPA 740-C-15-002 (2015), ISO 7346-1 (1996a), ISO 7346-2 (1996b), ISO 7346-3 (1996c)

(continued)

Table 4.1 (continued)

Organisms' category	Species	Toxicity endpoints	References
Amphibians	*Xenopus laevis*	Mortality, growth, development, length, histopathology, metamorphosis, reproductive maturity	OECD 231 (2009b), OECD 241 (2015)

seasonal variation of body fat (Oroş 2002). Fish species located at the top of the ecological pyramid, have been widely used as bioindicators of pollution of aquatic ecosystems with various toxic elements, including heavy metals. Having a regime of predators or higher predators, fish are likely to present BCF of 105–106, relative to the water content of certain mineral or organic contaminants (Oroş 2002).

4.5.2 Classification of Toxicological Tests Based on the Medium

Freshwater toxicity testing has considerably focused on liquid (complex environmental samples, chemical and biological contaminants) and solid media (sediments) assessment (Blaise and Férard 2006). Depending on the exposure medium, the tests are divided into (Rand 2008) water tests, whole effluent tests and tests in sediment.

The majority of aquatic toxicity tests are **water tests** (WT) by which the aquatic organisms are subjected to a single chemical or a chemical mixture spiked into water. The toxicity endpoints are comparatively evaluated in test organisms subjected to spiked water and non-spiked water (Rand 2008).

Whole effluent tests (WET) evaluate the aggregate toxic effects to aquatic organisms from all pollutants contained in a wastewater effluent compared to control sample. They can be performed in laboratory conditions or in situ. The objective of WT or WET is to estimate the "safe" or "no effect" concentration of these substances, which allow normal propagation of fish and other aquatic life in the receiving waters. In the case of the acute test, two water samples are required over a three-day (72 h) and for the chronic test three samples collected over a six-day (144 h) period (Adams and Rowland 2002; Rand 2008).

The 17 WET test methods for use in National Pollutant Discharge Elimination System (NPDES) monitoring (60 FR 53,529) measure the aggregate acute and chronic toxicity of an effluent using standardized freshwater, marine, and estuarine plants, invertebrates and vertebrates (EPA 1996). The species used must be sensitive to toxic substances, dominant in the general health of the food chain and representative of the possible impact area. The most relevant organisms are *P. promelas*, *C. dubia* and *Selenastrum capricornutum* which are used in the EPA 821-R-02-013 freshwater short-term toxicity test. For this test a control and a minimum of five effluent concentrations are assessed. According to the EPA recommendation to apply a 0.5 dilution factor, the effluent test concentrations of 6.25%, 12.5%, 25%, 50%, and 100% are frequently used. Toxicity endpoints are survival (NOEC), growth (IC25, IC50, NOEC) for *P. promelas*; survival (NOEC) and reproduction (IC25, IC50, NOEC) for *C. dubia*, and growth (IC25, IC50) for *S. capricornutum* (EPA 821-R-02-013).

To specify the contribution of each contaminant to the total toxic effect, toxicity units (TU) are used in ecotoxicity tests. The toxicity unit (TU = C/EC50) is obtained by converting the concentration of a contaminant taking into account its toxicity expressed by toxicity indicators (CL50) for reference organisms (*Daphnia*, CL50-48 h). Thus, the toxicity of a water sample will be quantified as the sum of toxicity units for all contaminants determined by chemical analysis.

Most of the chemical substances and their constituent elements, of anthropogenic or natural origin, eventually/ultimately end up accumulating in the sediments. **Sediment toxicity test** is performed in order to determine high concentrations of toxic chemicals that can cause damages at any level of aquatic organism or ecosystems (Burton et al. 2002). Micro- and macrobenthic species that live in or on sediments are exposed to various chemicals. Therefore, although sediment toxicity testing is a rather new field in ecotoxicological assessments, this research is important for the evaluation of toxicity, bioaccumulation and bioavailability of chemicals for benthic habitats.

The most used categories of organisms for these tests are the worms, clams, fish and phytoplankton (Lyubenova and Boteva 2016a, b). Both benthic species from fresh water, but also from the water column, such as *Selenastrum capriconutum, D. magna, C. dubia, P. promelas, H. azteca, C. tentans, Chironomus riparius, Hexagenia limbate* have been successfully used in sediment toxicity assessment studies (Borgmann et al. 2005; Burton et al. 2002). Although most sediment toxicity tests have been acute studies, currently increasing attention is being paid to chronic sediment toxicity tests with sensitive organisms and sensitive life stages. Full life cycle tests can be performed with freshwater mussels (*C. tentans* and *Paratanytarsus disimilis*)—EPS 1/RM/32 and with freshwater amphipods (*H. azteca*)—EPS 1/RM/33 (Adams and Rowland 2002).

According to Borgmann et al. (2005) the Sediment Quality Triad (SQT) is appropriate for assessing simultaneously sediment chemistry, toxicity and field alterations. The authors reported that the young specimens of *H. azteca* are the most suitable for evaluating the toxicity of whole sediments (ISO 16303: 2013), considering the sensitivity shown in terms of survival rates and growth inhibition after 14 and/or 28 days. Chapman (1990) described the SQT approach (a combination of chemical and ecotoxicological data, information on the biological community) as an important tool to determine pollution-induced degradation.

The nematode *Caenorhabditis elegans* stands out as another commonly used model organism in freshwater sediment toxicity tests (Heise et al. 2020). OECD Test No. 218 (2018) (spiking of the water column)/219 (spiking of the sediment) allow the evaluation of the effects of prolonged exposure of sediment-dwelling larvae of the freshwater dipteran *Chironomus sp.* to the chemicals. The duration of the test is 20–28 days for *C. riparius* and 28–65 days for *C. tentans.*

4.5.3 Classification of Toxicological Tests Based on the Exposure Timeframe

The exposure timeframe is an essential characteristic of exposure, determining different risks and outcomes. The time window of exposure should be selected according to the particularities of the study, such as the biological pathways of action of the potential toxicant, the test organism, the practical limitations of the study, etc.

Toxicity tests are generally divided into acute toxicity tests (high doses applied for a short time) or chronic toxicity tests (about 1/100 to 1/1000 of the acute dose applied for more than 3 months). Other derived tests can be performed as subacute toxicity tests (lower doses than those in acute tests applied repeatedly for 21–28 days) or subchronic toxicity tests (multiple exposure, for 1–3 months). Depending on the biological systems, media and toxicants exposure will be measured after a given period of time (Lyubenova and Boteva 2016a, b).

Acute toxicity tests are short-term tests (usually 24 to 96 h) by which the effects of a series of dilutions of the test chemical or the medium on the survival rate, growth and reproduction are monitored and recorded. Many studies have been devoted to finding the most sensitive species useful in acute toxicity tests (Gavrilescu 2008; Kennedy et al. 2002).

In environmental risk assessment, acute ecotoxicological testing is relevant on the one hand in screening a wide-ranging biological effect of the substance at the test doses and for the monitored indicators, and on the other hand in measuring the dose–response function and determining the LC_{50}/LE_{50}.

The acute toxicity test is usually carried out by performing five experimental variants defined by different doses (or concentrations) of the toxicant and a control without the toxicant. Each variant is performed in triplicate and the number of individuals in the tested sample must be large enough for the results to be processed statistically. For each test, a sample of individuals from the experimental species is used, which must be homogeneous and representative. The tests are carried out under standard conditions established by the methodological guidelines. The mortality results are recorded in tables and then the dose–effect (or concentration-effect) diagram is drawn establishing the LC-50 or EC-50 of the test compound in organisms. The acute toxicity tests are the first step for disclosing the total toxic effects generated by the toxicant.

Chronic toxicity tests are performed to assess the potential negative impact of a chemical after a long exposure timeframe on ecosystems (Gavrilescu 2008; Kennedy et al. 2002). Chronic ecotoxicological testing should cover entire life cycle or multiple generations or should cover several stages in the life cycle. Reproduction, growth, developmental effects and viability of the offspring are often evaluated (Kennedy et al. 2002). Chronic tests last for at least 10% of the tested species' lifespan; for-instance, 21-days of chronic exposure for invertebrates and more than 6 months for fish are common approaches (Rand 2008). Tests performed throughout the life cycle are expensive and long-lasting, especially if vertebrate species are used as test organisms (Lyubenova and Boteva 2016a, b). For this reason, many studies of the outcomes of chronic exposure to contaminants are executed on the most sensitive stages in the life cycle of fish, such as eggs, embryos, larvae and early stages of development. Biomarkers, such as enzyme production and activity are often approached for assessing the chronic toxicity of a chemical. Consistently, semi-chronic tests are carried out, following the effects during a sensitive period in the life cycle of the species, for example the reproduction period, the growth period, the migration period, etc. In chronic toxicity tests, aquatic organisms that can complete their entire life cycle in laboratory conditions, such as algae and invertebrates are particularly used. Evaluation of survival and fertility by age is carried out in accordance with Lotka's or Yuler's equation. On the whole, the chronic toxicity tests provide information allowing extrapolation of the effects of the pollutant agents at the community and ecosystem level (Lyubenova and Kalchev 2009).

4.5.4 Classification of Toxicological Tests Based on the Mode of Exposure to the Tested Substance

In a **static-toxicity approach**, the test organisms are exposed from the beginning to the test solution, which is not renewed over the exposure timeframe. The static tests are simple and cost-effective and are generally performed in short-term exposure ($\leq$ 96 h with fixed or slowly degradable materials and at a low load of the test organisms). Static systems are performed when the critical load of toxic has a insubstantially availability with wastewater (Lyubenova and Boteva 2016a, b);

Recirculation tests are similar to static ones, but the test solutions are pumped through a filter to maintain water quality, without changing the test concentrations of the chemical.

Semi-static tests (renewal) imply periodic renewal of the test solution, usually once a day. The renewal systems are applied when the test organisms are sensitive to streams or when, due to their small size, they can be lost in the watercourse systems. A semi-static toxicity approach is applied when the test material is limited (Lyubenova and Boteva 2016a, b). The exposure time varies by species, for example between 7 and 21 days in *Daphnia* sp. tests, between 7–10 days in juvenile fish tests.

In continuous flow tests the water and the toxicant are renewed continuously or intermittently. These are the most suitable because partial loss of the toxic by evaporation, by fixation or metabolism or by precipitation is avoided. In the acute toxicity study of substances with high volatility and those that are not persistent in water, flow-through systems are recommended.

Irrespective of the mode of exposure, all kind of tests use a water control for comparison of the outcomes.

4.5.5 Classification of Toxicological Tests Based on the Number of Test Species

Adverse effects of chemicals from molecular to individual level and to populations and communities are commonly evaluated in the environmental risk assessment. The study of exposure pathways and response induced by a stressor is often focused at the individual level, especially as a starting point in ecotoxicological studies (Petre and Teodorescu 2009). The **single-species test** is far less labour- and time-consuming than experiments on higher levels of biological organization and are, therefore, a fundamental tool in the hazard identification and potential harmful impacts of various pollutants on aquatic life. These tests ensure a high degree of replicability and provide relatively accurate outcomes of the toxicity endpoints under standardized conditions (Breitholtz et al. 2006).

The vast majority of toxicological data are in the form of EC50 and LC50, generated by single-species toxicity tests (Lyubenova and Boteva 2016a, b). In the European Union, a safety environmental factor of 100 is applied for acute toxicity tests using the test organisms *D. magna* and fish (EEC Council Directive 1991).

Monospecific testing refers, first of all, to acute toxicity tests, which are applied for a limited duration in relation to the life span of a biological individual. Single-species tests are suitable for determining toxic effects at the individual level: mortality, growth, reproductive capacity, behavior (Petre and Teodorescu 2009). However, the outcomes are not sufficient to extrapolate responses to other species.

The results obtained from single-species toxicity tests are not applicable for predicting the behavior of integrated species in complex freshwater ecosystems. Biological effects within the natural freshwater system with variable environmental conditions differ considerably from standard laboratory tests (Lyubenova and Kalchev 2009). For example, populations are generally less sensitive to pollutants than individuals. The generalization of the toxic effects found in laboratory conditions to the level of the natural ecosystem requires an excellent knowledge of its structure and functional processes. Cumulative risk assessment, bioaccumulation and biomagnification of a contaminant through transfer within aquatic food webs is mandatory for the correct understanding of its relevance (Lescord et al. 2015).

Taking into account these impediments, a more realistic approach to the problem can be represented by monitoring the effects of a potential harmful factor in natural ecosystems or in specially man-designed systems to approximate the overall characteristics of a habitat, such as **multi-species tests** with the key representatives of studied ecosystem.

An optimal combination of test organisms for aquatic ecosystems (reflecting their functional structure) includes algae, crustaceans and fish. Although these microecosystems are not completely analogous to natural ones, they are sufficiently complex to allow realistic ecological study of the typical characteristics of ecosystem in nature, as well as data extrapolation (Lyubenova and Kalchev 2009).

The Taub test is a classic example of a standardized multi-species test performed with 24 identical containers colonized with ten species of algae, five species of animals and a set of unidentified bacteria. After 60 days, community deterioration occurred in the absence of the normal biological cycle. In another classic test designed by Kersting, a microcosm with 151 compartments was created. Primary and secondary production, and decomposition were separated to prevent overconsumption. The test was a success, with the balance between production and decay proving sustainable for months or even years under relatively stable conditions, corresponding to "ecological time periods". Another well-known test was proposed by Cairns and Lundgren for studying interactions between algal and water fleas' populations under chemical stress (Lyubenova and Kalchev 2009).

Based on the different sensitivity of organisms, Dutka et al. (1988) emphasized the importance of applying "bioassay batteries" (combinations of bioassays), claiming the need for more species from different trophic levels, in assessing sediment quality. In order to assess the acute toxicity of water-extractable fractions of biochar-amended soil, Bastos et al. (2014) conducted one battery of bioassays with representative aquatic organisms, such as *Vibrio ęscheri,* microalgae *P. subcapitata* and the invertebrate *D. magna.* In another battery of bioassays *C. reinhardtii, D. magna, C. riparius* and *D. rerio* were used to assess the impact of leak of effluent from an Installation of Cleansing and Uranium Recovery (Tricastin, France) along a river (Gagnaire et al. 2011).

A number of studies have used aquatic ecosystem models designed to simulate ecosystems or parts of natural ecosystems. Artificial ecosystem models are relatively easy to handle logistically and statistically (Odum 1984).

Multi-species ecotoxicity tests vary in both size and complexity. The most used experimental method is ecotoxicological testing in replicated artificial ecosystems. Depending on their size, artificial ecosystems are differentiated into macrocosms, mesocosms and microcosms (Caquet et al. 1996). A microcosm is an artificial and controlled simplified ecosystem, made indoors (fish aquarium) or outdoors (small ponds), in volumes less than 1000 L, and consisting of two or more species. Mesocosms represent a complex ecosystem (mimic natural conditions) made up of several species placed in experimental volume of water that exceed 1000 L and the test is usually done outdoors. Macrocosms are the most realistic (lakes and on whole aquatic ecosystems) but they are very difficult and expensive to conduct (Rand 2008).

Artificial systems are relatively variable over time because they are species-poor and lack redundancy (Wong et al. 2004). Also, they are poorly replicated and this raises concerns about the accuracy of the prediction.

Mesocosms can be designed to simulate both lentic (ponds) and lotic (streams) freshwater ecosystems. After stabilization, sediments, plants, and animals from natural ecosystems are introduced (Caquet et al. 1996; Rodgers et al. 1996). Mesocosms offer the advantage of controlling and monitoring multiple factors such as exposure time, duration, and concentration, and statistical interpretation of results is supported by similar replication units.

Studies of aquatic mesocosms are still rare, although indispensable, because modest effects manifested at a single trophic level can branch and amplify through the ecological network, influencing the function of the entire ecosystem (Jabiol et al. 2013). Mesocosm studies have also reported conflicting results regarding the development of primary production (Fairchild et al. 1992), probably due to different ecological interactions between phytoplankton, zooplankton, and fish.

A mesocosm study was conducted by Bour et al. (2016) to investigate the toxicity of cerium oxide nanoparticles (CeO_2NPs) on an experimental freshwater ecosystem consisting of diatoms (*Nitzschia palea* and *N. pelliculosa*), the macroinvertebrate chironomid (*C. riparius*) and larvae of the carnivorous amphibian species *Pleurodeles waltl* (the highest trophic level in that ecosystem). The species *C. riparius* is representative of many aquatic ecosystems and ecotoxicological tests. Chironomid larvae have significant interventions in processing sediments and biofilms, thus making them potential vectors for transfer in food chains.

4.5.6 Classification of Ecotoxicological Tests Based on the Place of Performance

Most **ecotoxicological tests are carried out in the laboratory** and others **under natural conditions**. Samples can be collected from the field and later analyzed in the laboratory, especially when the purpose of the test is the research of some effluents.

Alternatively, toxicity tests can take place in situ, which means exposing the test organisms to real environmental conditions (temperature, flow rate, light), the data thus obtained providing a more realistic assessment of toxicity than data obtained from laboratory analyses (Birge and Black 1981; Castaño-Sánchez et al. 2021). For in situ aquatic toxicity tests, fish, clams and oysters are used.

On the other hand, in situ tests also have certain disadvantages: lack of control over the conditions in which the test takes place, data interpretation must take into account how variable conditions could affect the results, and last but not least, logistics, which can prove difficult in extreme weather conditions. In these tests, it is recommended to use native species relevant to the ecosystems under investigation, which are adapted to the specific environmental conditions. The use of native species presents the advantage of preventing the introduction of exotic or invasive species into the environment, but it also presents the disadvantages of an unknown life history and the difficulty of evaluating some traits, such as the age of individuals, before exposure to stressors (Castaño-Sánchez et al. 2021).

4.6 Ecosystem Modelling

Ecological bioaccumulation models have been proposed to assess the bioaccumulation potential of chemicals, to develop new water and sediment quality assessment criteria, and to measure the impact of pollution on biota. However, because these models require species-specific toxicodynamic and toxicokinetic data, they are not widely applicable.

From the first bioaccumulation model proposed by Hamelink et al. (1971), various bioaccumulation assessment methods have been proposed. Testing a single concentration is considered sufficient when the bioconcentration factor (BCF) is independent of the concentration to be tested. The European Center for Ecotoxicology and Toxicology of Chemicals (ECETOC) proposed a model of an integrated bioaccumulation testing strategy, which involves three stages of bioconcentration assessment: simulations, non-animal experimental systems and testing on a decresed number of animals relatively to OECD Test No. 305 (2012b), which is reserved for the final stage of the strategy (ECETOC 2007). ECETOC analyzed several case studies evaluating the aquatic toxicity and bioaccumulation of several chemicals considering the "mode of action" (i.e., the set of physiological and behavioral responses induced), and developed four subcategories, depending on the type of protein interaction: receptor, ion, channel, enzyme or transporter. A weight-of-evidence (WOE) approach similar to this integrated testing strategy is included in the ECHA guidance for REACH (ECHA 2008).

Physiological models for investigating bioconcentration and bioaccumulation in fish were developed by Barber et al. (1988, 1991, 2008); Law et al. (1991); Nichols et al. (1990). The role of diet, gill absorption and biomagnification in the bioaccumulation of hydrophobic organic contaminants has been studied by Bruggeman et al. (1981); Clark et al. (1990); Connolly and Pedersen (1988); Muir and Yarechewski (1988); Niimi and Oliver (1987) and others. Food web models were developed by Arnot and Gobas (2004); Gobas (1993); Thomann et al. (1989, 1992).

The study of the transfer of chemicals from sediments to benthic invertebrates has allowed the development of bioaccumulation models, which vary from the study of equilibrium partition (DiToro et al. 1991) to detailed kinetic models (Landrum et al. 1992; Morrison et al. 1996).

Structure activity relationships (SARs) relates physical or chemical features of a chemical structure or substructure with a certain biological activity/effect. When the result is expressed quantitatively, the relationship is a quantitative structure activity relationship (QSAR).

QSAR models are a non-testing method to predict acute or chronic toxicity data (Chen and Schüürmann 1997; Netzeva et al. 2008). A report presented by the European Chemicals Agency showed that the use of QSAR models can be applied to predict bioaccumulation, as well as short- and long-term toxicity in fish (ECA 2014). EPA has routinely applied QSAR approaches to new chemical registrations since the 1980s (Nabholz et al. 1993). Starting from the structure of compounds with known toxicity, QSARs have been used to predict the toxicity of molecules that have not been conventionally tested.

While standard toxicity tests use constant or pulsed exposure, Ashauer et al. (2007) successfully integrated unsteady and periodic concentrations of a pesticide using a threshold damage model to predict the effects of realistic exposure conditions on the invertebrate *G. pulex*.

Individual-level models can be used to analyze acute toxicity test results and explore mechanisms. For example, the energy budget model describes characteristics of individuals such as growth, metabolism or reproduction (Nisbet et al. 2000). The most frequently used energy budget model in ecotoxicology is DEBtox (Jager 2020; http://www.bio.vu.nl/thb/deb/deblab/debtox/).

Species sensitivity distributions (SSDs) were introduced to generate more precise environmental quality targets (Kooijman 1987). SSDs integrate the results of single-species assays for a particular compound (or mixture of compounds) to establish a statistical distribution of sensitivity. However, the effects of toxicants in freshwater ecosystems are addressed by only a few mechanistic models, which have rarely been applied in ecotoxicology, are not fully validated and compared with field data (Sourisseau et al. 2008). The Aquatox ecosystem model (http://www.epa.gov/waterscience/models/aquatox/) for control flows in a mesocosm experiment was calibrated and validated by Sourisseau et al. (2008). The model is characterized by sensitivity to the optimal growth temperature for periphyton, filamentous algae and predatory invertebrates.

Statistical models that extrapolate concentration–response relationships from case studies or from lower levels of biological organization to a broad across-level scale are an alternative to mechanistic effect models. For example, Schriever et al. (2007) combined a mechanistic fate model with a statistical model including the observed dose–effect relationship in two central European regions to predict pesticide effects at the European level.

Arnot and Gobas (2004) designed a model of partitioned bioaccumulation of hydrophobic organic chemicals within food webs in algae, phytoplankton and zooplankton. The model aims to predict concentrations, brachial ventilation rates based on allometric relationships and a mechanistic model for predicting gastrointestinal magnification of organic chemicals in a range of aquatic species.

4.7 Plant-Based Ecotoxicological Tests

Being at the base of most food chains, plants are the first to accumulate and then introduce contaminants into the food chain, so various phytotoxicity tests have been developed by international organizations such as ISO, OECD, EPA, etc.

The ECOTOX database indicates the use of over 6000 species of aquatic organisms in ecotoxicology, of which only 25% are plant organisms (microalgae, flowering plants, macroalgae, pteridophytes, and aquatic bryophytes (Ceschin et al. 2021; EPA 2023). Among the most used species are the microalgae *P. subcapitata*, *Chlorella vulgaris*, *Chlamydomonas reinhardtii*, *Chlorella pyrenoidosa*, and *Scenedesmus quadricauda*, the macroalgae *Cladophora glomerata*, *Oedogonium cardiacum*, *Spirogyra* sp., *Chara* sp., and *Zygnema* sp., the Bryophytes *Fontinalis antipyretica*, *Platyhypnidium riparoides*, *Vesicularia dubyana*, *Scapania undulata*, and *Fontinalis dalecarlica*, the Pteridophytes *Azolla pinnata*, *Salvina molesta*, *Azolla caroliniana*, *Salvinia natans*, and *Azolla filiculoides* and the flowering plants *Lemna minor*, *Myriophyllum spicatum*, *Hydrilla verticillata*, *Lemna gibba* and *Myriophyllum sibiricum* (Ceschin et al. 2021). *Spirogyra* sp., *C. pyrenoidosa*, and *L. minor* have been mentioned in ecotoxicological research since 1931, 1952, and 1954, respectively.

Phytotoxicity evaluation with various plant species is absolutely necessary, the sensitivity to xenobiotics present in their environment varying considerably (Van den Berg et al. 2020). For example, comparative tests carried out with *L. gibba* and *L. minor* showed that *L. gibba* was more sensitive to atrazine, diuron, irgarol, and pentachlorophenol, whereas *L. minor* was more sensitive to Cd, Cu, and triclosan (Lewis and Thursby 2018).

The OECD Test No. 201 (2011a) is based on exposure to xenobiotics of some species of cyanobacteria or microalgae for 72 h. During the test period, the algicidal or algistatic effects can be evaluated, quantified by determining the algal biomass. Statistical analysis of the results provides reliable information about the inhibition of growth rate or yield. Furthermore, NOEC and LOEC can be determined (Table 4.2).

The algae and cyanobacteria species recommended for performing the toxicity test are the green algae *P. subcapitata* and *D. subspicatus*, the diatoms *N. pelliculosa*, and the cyanobacteria *A. flos-aquae*, and *S. leopoliensis*, the most appropriate being the green alga *P. subcapitata*. Querying the Web of Science database by species name, refined with the keyword toxicity, yielded approximately 852 publications for *P. subcapitata*, 99 for *D. subspicatus*, 18 for *N. pelliculosa*, 151 for *A. flos-aquae*,

Table 4.2 Measured parameters and phytotoxicity indicators in standardized tests

Test/species	Exposure period	Measurement parameters	Toxicity indicators
OECD 201 (2014a)/ microalgae and cianobacteria	72 h	– growth and growth inhibition – cell volume – fluorescence – optical density, etc.	ECx, NOEC, LOEC
OECD 221 (2006)/ *Lemna gibba*, *L. minor*	7 days	– frond number – total frond area – dry or fresh weights	ECx, NOEC, LOEC
OECD 239 (2014b)/ *Myriophyllum spicatum*	14 day	– main shoot length, side shoot number and side – visual assessments of plant health – shoot fresh and dry weights – visual assessments of root health	ECx, NOEC, LOEC

and 14 for *S. leopoliensis*. For these species, the most recent research practices have evaluated the toxicity and ecological risk (Udebuani et al. 2023) of nanoparticles (Mahaye and Musee 2023), antibiotics (Li et al. 2023), veterinary pharmaceuticals (Udebuani et al. 2023), antihistamine drugs (Roveri et al. 2023), pesticides combined with toxigenic cyanobacteria (Hernández-García and Martínez-Jerónimo 2023), etc. on apical growth and molecular mechanisms (Li et al. 2023; Mahaye and Musee 2023). In silico methods are currently applied in predictive research related to the fate of tested xenobiotics in the environment (Roveri et al. 2023).

Another standardized phytotoxicity test recommended by the OECD TG 239 (2014a) is based on the use of the species *M. spicatum* and the water–sediment method. The roots that characterize the submerged dicotyledonous species *M. spicatum* L. (Eurasian milfoil) (family Haloragaceae) (URL 1) recommend it for studies in which other organisms are not sensitive to substances that accumulate in the sediment, the exposure being through root absorption. After appropriate modification of the methodology (test conditions, exposure timeframe, etc.), other species of aquatic plants, such as *Myriophyllum aquaticum* and *Glyceria maxima*, can also be used for the testing proposed by the OECD Test Number 239 (2014a) through the water–sediment method. The application of this test leads to the establishment of the route of exposure to the test substance (i.e., via water or sediment).

M. spicatum plants required for testing can be harvested from the environment or grown in greenhouses or laboratories. The plants in vegetative phase must be kept free from any biological contaminant, and acclimatized for a period of > 2 weeks.

Cultivation of plants is done in an artificial sediment with a pH of 7 ± 0.5, which contains peat (4–5%), kaolin clay (20%), quartz sand (75–76%), and nutrients (ammonium chloride and sodium phosphate). An inert material is distributed on the surface of the sediment. The liquid phase was Smart and Barko medium at pH 7.5–8.0 (Smart and Barko 1985).

The biological parameters evaluated during the test period are presented in Table 4.2. The recorded data allow the calculation of average specific growth rate (r) and yield (y), which illustrate the influence of the tested substance on the vegetative growth of *M. spicatum*. The expression of ErCx and EyCx ($x = 10, 20, 50.$) is based on the results obtained for average specific growth rate (r) and yield (y) growth rate (r) and yield (y). The statistical analysis of the results allows the calculation of NOEC and LOEC.

Species from the genus *Lemna* are among the most used for toxicity assessment, as it appears from the scientific publications of the last two decades, visible in Web of Science (*Lemna*/toxicity: 1001 publications) (URL2). Including in the ECOTOX database, the species *L. minor* records the highest percent (40.12%) of publications for pteridophytes and flowering plants (Ceschin et al. 2021).

The OECD Test No. 221 test is based on the use of *L. gibba* and *L. minor* species in the exponential growth phase, with the xenobiotic exposure period of 7 days. Frond number, total frond area, and dry and fresh weight are the biological parameters measured. Toxicity tests based on *L. minor* have also been carried out by other organizations such as: Environment Canada (1998), (URL3), ISO (ISO DIS 20079 2019), Association Française de Normalisation (AFNOR 1996) etc. The latest ecotoxicological research that involved *Lemna* species, has provided relevant information regarding the toxicity of microplastic (Cui et al. 2023), zinc (Trenfield et al. 2023), difenoconazole and atrazine (Mendieta Herrera et al. 2023), effects of gamma radiation and zinc (Bodnar and Cheban 2023), tire wear particle toxicity (Roubeau Dumont et al. 2023), lanthanide-doped nanoparticles toxicity (Ekner-Grzyb et al. 2023).

In recent years, several aquatic macrophytes and other toxicity target organisms have been considered for environmental risk assessment (Sesin et al. 2023). The investigation of the Web of Science database shows that in the last two decades (2003–2023), the species *M. spicatum* was mentioned in 108 papers (after the keywords *Myriophyllum spicatum*/toxicity). Among other objectives, research aimed at improving experimental designs and quantify their benefits by predicting growth responses (Heine et al. 2014), the bioaccumulation capacity of heavy metals (Chowdhury et al. 2023), mesocosm-type experiments (Mohr et al. 2023; Zingel et al. 2023), evaluating the impact of some metal nanoparticles (Roubeau Dumont et al. 2022) or micro- and nanoplastic (van Weert et al. 2019; Xie et al. 2023).

Together with species of *Ceratophyllum, Elodea, Egeria, Azolla,* and *Pontederia, Myriophyllum* is part of "*Green Liver Systems—systems with aquatic macrophytes that have been developed as low-impact, low-energy, and low-cost solutions for the remediation of water pollutants*" (Esterhuizen and Pflugmacher 2023).

4.8 Ecotoxicological Tests on Freshwater Invertebrates

Compared to other groups, the higher frequency of macroinvertebrates in monitoring programs is justified by the large number of species, detailed taxonomic description, their sedentary nature and low-cost methods (Rosenberg and Resh 1993).

The role of ciliates in aquatic ecosystems, including in water purification processes, and their sensitivity to changes in the environment, made scientists interested in the use of these unicellular protozoan for aquatic toxicology studies where the effect of polluting factors on lakes, ponds, bodies of stagnant water, rivers, etc. is investigated (Apostol 1973; Gilron and Lynn 1998; Komala 1995; Nicolau et al. 1999). With the increasing understanding in the potential use of ciliates as a tool for toxicological studies, study methods were proposed in the middle of the twentieth century, adapted to the characteristics of the species and to technological knowledge since that time (Persoone and Dive 1978).

There is a permanent concern for the development of frugal and viable standardized alternative methods to conventional ones, although many of the recent methods are based on classical criteria, namely taxonomic identification and mortality of organisms exposed to pollutants (Janssen et al. 2000). Its ubiquity, the large size of cell, the ease in obtaining cultures, and the rapid rate of division are attributes that led many researchers to focus on paramecia as an ideal model for investigating water quality fluctuations and risk assessment of chemical pollution. Among them are of notice *Paramecium caudatum, P. primaurelia, P. bursaria* (Amanchi 2010; Apostol

1973; Mills 1976; Miyoshi et al. 2003; Rao et al. 2006; Zhang et al. 2022). These research papers were preceded by studies on biology and appropriate methodologies for obtaining paramecia cultures, the works of Sonneborn (1970) and Wichterman (1986) being the reference in this field.

The toxicity evaluation targets morphological, ultrastructural, ethological (mobility, stress, nutritional behavior), and metabolic (growth and reproduction) criteria (Amanchi 2010; Amaroli et al. 2010; Persoone and Dive 1978; Rao et al. 2006; Takiguchi et al. 2002).

The classical procedure to evaluate chemical pollution consists in exposing paramecia to the half-maximal inhibitory concentration (IC50) of the target pollutant under laboratory conditions. This is then followed by microscopic observations on the effect of the pollutants on paramecia reproducibility/mortality (Mills 1976; Miyoshi et al. 2003). In the last decades, the cytotoxic effect is associated with morphological and functional alterations, emphasized also due to the advances made in the field of molecular biology and border disciplines.

Rao et al. (2006, 2007), using conventional microscopy and computerized video tracking systems, investigates the morphology, generation time, and locomotor behavior of the species *P. caudatum*, by exposing paramecia cultures to LC50 and sub-lethal concentrations of acephate and monocrotophos, two organophosphorus insecticides. Shunmugam et al. (2021) found a significant relationship between the swimming speed of paramecia and viscosity, using a culture medium modified by adding some categories of pollutants, such as heavy metals, antibiotics, organic solvents. The simple, automated procedure requires only a few minutes of video recording compared to a few days for classic IC50 measurements and can be implemented on a global scale, regardless of the level of development of the region. In the past, the measurement of some parameters of ciliates, such as motility, depended on the abilities of researchers to use the research technique of that time. Bergquist and Bovee (1974) combined Polaroid equipment with stroboscopic discs and projection of negatives on a large screen to capture the movement of ciliates. Nowadays, high-performance equipment (most of its computer-assisted), video recordings, etc. ease the work of specialists in this field, the new generation of technologies being able to cause a real revolution in ecology studies, even in aquatic toxicology (Allan et al. 2018; Ohmura et al. 2018). However, the economic component of modern methodologies should not be neglected in their need to be implemented in underdeveloped areas of the world, considering the significant challenges that human society will face in the coming decades in the fields of energy, food, health (Prabhu 2017). The use of ciliates as an experimental model is thus added to another group of experimental aquatic organisms, fish, the biomarkers most used in environmental risk assessment (van der Oost et al. 2003).

The toxicological tests performed on daphnia aim at evaluating the degree of immobilization (acute toxicity test) or reproductive impairment (chronic toxicity tests).

Thus, the OECD Test No. 202 (OECD 2004) and ISO 6341 (ISO 2012) tests are acute toxicology tests by which the concentration of a substance that immobilizes daphnia species is determined. Thus, individuals are exposed to a range of concentrations of the test substance for 48 h, and immobilization is recorded at 24 h and 48 h (EC50 is determined) and compared to control values.

OECD Test No. 211 (2012c) determines the effect of a chemical on the reproduction and mobilization of *D. magna*. The test involves exposing female *D. magna* (less than 24 h old at the start of the test) to a range of concentrations of the test substance. The duration of the test is 21 days and the total number of live offspring produced is recorded.

A chronic toxicity test on daphnia with a maximum age of 1 day (*Daphnia magna* or *Daphnia pulex*) was proposed by the US EPA-OCSPP 850.1300 Daphnid Chronic Toxicity Test (201). The 21-day test monitors survival, growth and reproduction, and the results are used to calculate the NOEC and LOEC for these endpoints.

There are several tests, such as EPA-HQ-OPPT-2009-0154-0040, OCSPP 850.1020 carried out on immature gammarids (*G. fasciatus*, *G. pseudolimnaeus* or *G. lacustris*) exposed to the test substance for 96 h, during which observations are made on survival rate and other toxic effects. The test is designed to determine the relationship between the test substance and mortality of gammarids over the full concentration–response curve (96 h LC50).

In the Standard Guide for Conducting Laboratory Toxicity Tests with Freshwater Mussels (ASTM E2455-06 2013) describes methods of using early life stages of freshwater mussels, including glochidia, and juvenile mussels in water-only exposures.

Tests on chironomid larvae (Diptera), such as *C. riparius*, *C. dilutus* or *Chironomus yoshimatsui* can be performed using spiked water (OECD Test No. 219) or spiked sediment (OECD Test No. 218) to study the effect of the increased levels of the chemicals into their sediment-dwelling. The duration of the tests is 20–28 days for *C. riparius* and *C. yoshimatsui* and 28–65 days for *C. dilutus*. The duration of the tests is 20–28 days for *C. riparius* and *C. yoshimatsui* and 28–65 days for *C. dilutus*. The number of emerged males and females, the time to emergence and the larval survival and growth after 10 days of exposure are determined. The exposure of the previously mentioned chironomids to the chemicals in the sediment can be performed over the whole life cycle (OECD Test No. 233). In this case the exposure timeframe increases to 44 days for *C. riparius* and *C. yoshimatsui* and about 100 days for *C. dilutus*; the emerged adults are transferred to breeding cage and the number of the egg's produce and their fertility are assessed. Likewise, the viability of the second generation is determined. The three chironomid species can be used in immobilization tests similar to those for daphnia (OECD Test No. 235).

Freshwater snails are also used in ecotoxicological investigations. In the OECD Test No. 243 (2016b) toxicity test, after exposure of reproducing adults of *L. stagnalis* for 28 days to different concentrations of chemicals, survival, reproductive endpoints (number of eggs) and shell length of the adults are determined. The OECD Test No. 242 (2016a) test assesses survival and embryo numbers in the brood pouch of the freshwater snail species *P. antipodarum* exposed to chemicals for the same period of time.

The endobenthic oligochaete *L. variegatus* plays an important role in the sediments of aquatic ecosystems through bioturbation and as prey species. OECD Test No. 225 (2007) determines the effects of prolonged exposure (28 days) under static conditions of specimens of *L. variegatus* to the test substance on their reproduction and biomass. The test determines the EC50, as well as the NOEC and LOEC for biomass and reproduction, if supporting data are available.

Due to the growing demand for cost-effective tests performed at different levels of biological organization, small-scale toxicity tests (microbiotests) have found an increasingly wide applicability. Defined as "the exposure of a small unicellular or multicellular organism to a liquid/solid sample to measure a specific effect" (Blaise and Férard 2006), microbiological tests have demonstrated effective screening for potential toxicity. The development of rapid tests has led to a considerable increase in the number of studies on the applicability, reliability, reproducibility and relevance of these tests. However, rapid tests created several controversies, with some authors appreciating them as ineffective alternatives to traditional tests (Mitchell et al. 2002).

Because of their culture-free nature, rapid, user-friendly and space-saving characteristics, Toxkit's microbiotests represent a convenient option compared to classic ecotoxicological tests (Fochtman 2000; Tofan et al. 2023). These tests are initiated from dormant eggs or immobilized stages, which can be hatched or de-hatched by request, and are thus independent of the supply, cultivation, and/or maintenance of live stocks of test biota (Persoone et al. 2009). Microbiotests are very useful in acute ecotoxicological investigations for the pre-screening and screening of chemicals and wastewaters (Ruck et al. 2000).

As ecotoxicological tests used in researching the influence of toxic substances on freshwater organisms, the following can be used:

- Daphtoxkit F—the *Daphnia* immobilization test highly standardized, in compliance with ISO Standard 6341 and OECD Guideline 202. Outcomes: EC50 at 24 h and 48 h;
- Thamnotoxkit F—acute 24 mobility inhibition tests with the freshwater crustacean *Thamnocephalus platyurus*; highly standardized bioassay complies with ISO Standard 14380 (ISO 2011). Outcomes: LC50 at 24 h;
- Ceriodaphtoxkit F—acute toxicity tests with the freshwater crustacean *Ceriodaphnia dubia* are included in the kit as dormant eggs (ephippia) which can be easily hatched on demand in 72 h. The Ceriodaphtoxkit F adheres to the protocols for regulatory testing prescribed by international organisations; US EPA Whole E uent Toxicity Method 2002.0 (EPA 2002a, b, c, d, e);
- Rotoxkit F—acute 24 h mortality tests with the freshwater rotifer *Brachionus calyciflorus*; highly standardized bioassay in compliance with ISO Standard 19827. The test organisms are included in the kit as cysts (dormant eggs) which can be easily hatched on demand in less than 20 h;
- Rotoxkit F—chronic 48 h reproduction tests with the freshwater rotifer *Brachionus calyciflorus*, cyst-based rotifer toxicity; highly standardized bioassay in compliance with ISO Standard 20,666 (ISO 2008);

- Protoxkit F—chronic 24 h growth inhibition tests with the freshwater ciliated protozoan *Tetrahymena thermophila.* The test organisms are included in the kit in a steady state condition. Outcomes: growth inhibition NOEC of protozoan microfauna in benthic communities.

4.9 Ecotoxicological Tests on Freshwater Fish

Under the auspices of the Water Framework Directive (EU 2000), fish are used to improve the potential for assessing the quality of effluents and surface waters. Acute toxicity tests on fish for environmental hazard and risk assessment are among the top individual preferences of researchers (Braunbeck et al. 2015).

In addition to its economic importance, fish, as an adaptable test organism and as a suitable general model for vertebrates, is frequently used for the development and continuous improvement of test methods. According to Hinton et al. (2005), small fish represent an "ecological sentinel", a tool for detailed and rapid toxicity investigation. However, to design a reliable test it is necessary to know the behavior of the substance in the test system (physico-chemical properties and fate characteristics).

Pimephales promelas, also known as fathead minnow, is a sentinel species often used to assess the acute and chronic toxicity of chemicals and effluents (Ankley and Villeneuve 2006). Fathead minnow is one of the dominant models in aquatic ecotoxicology (Saari et al. 2016) finding itself in a wide variety of research applications such as assessment of acute and chronic toxicity of a chemical or mixture, QSAR, interspecific or population level extrapolation and to relevant field conditions of toxicological data for various regulatory applications, from basic 48/96 h lethality tests to complex partial and full life cycle tests (Ankley and Villeneuve 2006; Alsaadi et al. 2018; Erickson et al. 2022).

Oryzias latipes (Japanese medaka) is a teleost vertebrate model with a history of over 60 years in toxicity testing. Having a short life cycle and due to the possibility of genetic sex determination, embryos or adults of this small species are mainly used in short-term reproduction assay (Padilla et al. 2009; Onishi et al. 2021).

Danio rerio (zebra fish) is a well-established model for toxicity screenings, having a number of advantages: it adapts to variable environmental factors (temperature, pH), it has transparent embryos, which allows easy monitoring of toxic effects such as teratogenicity, lethality, reproduction (Bai and Tang 2020), has rapid development, high fecundity rate, gene manipulation is accessible, tests are cost effective, small size of adults (3–5 cm) allows easy maintenance and require small amounts of investigated chemicals (Hill et al. 2005). Furthermore, the nervous, cardiovascular, and digestive systems have similar functioning to that of humans (Hsu et al. 2007).

One of the widely used tests to assess the acute toxicity of chemicals to various fish species is OECD Test No. 203 (OECD 2019), recommended as a preliminary test. Standard testing involves exposing fish to a test substance for 96 h and recording mortality at 24, 48, 72 and 96 h, with LD50 determination possible. One of the most widely used species in static, semi-static and continuous flow acute lethal toxicity tests is *Brachydanio rerio* (ISO 7346-1, ISO 7346-2, ISO 7346-3). However, testing can be performed against a variety of method-approved fish species.

The species-specificity of the response to the same exposure can act at different life stages (eggs, embryos, larvae, adults). Early life stages of fish (embryos and larvae), recognized as a robust model for predicting toxic effects over the complete life cycle (Norberg-King 1989), have a number of advantages over tests performed on adults of the same species: they are less expensive, they allow the study of a larger number of species, the results can be extrapolated to several species characteristics of various areas and trophic levels (Lyubenova and Boteva 2016a, b).

Moreover, the developmental stages of fish that are not fed are not classified as protected stages under the new European directive 2010/63/EU on the protection of animals used for scientific purposes. Thus, fish embryos are used as surrogate for juvenile fish, being considered in many jurisdictions as a nonprotected life. In Germany, the 48-h fish exposure test (DIN 38415-6: 2003) for the assessment of acute effluent toxicity has been replaced by the 48-h post-fertilization zebrafish embryo exposure test (ISO 15088: 2007).

Attempts to reduce the duration of standard toxicological protocols without compromising their predictive value have been supported by the intention to decrease the costs of toxicity assessment and hazard identification tests. Thus, eight-day embryo-larval survival tests for fish and other aquatic vertebrates (Birge et al. 1985) and a seven-day larval survival and growth test (Norberg and Mount 1985) have been developed.

The OECD 212 test is a short-term toxicity test conducted over all life stages, from the newly fertilized egg to the end of the sac-fry stage, which are exposed to 5 different concentrations of the test substance. In order to define the lethal and, to a lesser extent, sublethal effects of chemicals on the specific stages and species tested. In this test, the survival rate, hatching percentage, appearance, behavior and days until hatching are also evaluated and recorded.

Originally designed as a replacement for the acute test (OECD Test No. 203), the zebrafish embryo test (FET) has been optimized, standardized, validated and adopted as OECD Test No. 236 (2013b). Performed on newly fertilized zebrafish eggs, lasting 96 of hours, the OECD Test No. 236 test monitors, every 24 h, the viability of the eggs: coagulation of fertilized eggs, lack of somite formation, lack of detachment of the tail-bud from the yolk sac and lack of heartbeat. The conclusion of an acute toxicity is formulated on the basis of positive results, for any of the four apical observations recorded, as well as an endocrine disrupting effect. Although OECD 236 provides reliable indications of acute lethality, its effectiveness has been questioned, with objections related to the alleged lack of biotransformation capacity and the assumption that highly lipophilic and/or high molecular weight substances may not have access to the embryo (Braunbeck et al. 2015).

Chronic toxicity tests in fish record and interpret data on the alteration of growth, reproduction and survival, but also the teratogenic potential of the stressor. The main parameters determined are NOEC, LOEC, LC_{50} and EC_{50}, as well as the abortion rate of eggs and embryonic stages, the decrease offspring, egg production, fecundity and fertility, reduced hatching success, failure of metamorphosis and developmental abnormalities. Lillicrap et al. (2016) appreciate that chronic toxicity tests, from subchronic effluent tests to full life cycle tests, are the most resource-intensive with respect to animal use compared to the other aquatic tests.

A first life cycle toxicological study in fish was done by Mount and Stephan (1967). The authors exposed individuals of *P. promelas* (fathead minnows) to a graded series of pesticide concentrations throughout their life cycle and assessed the effects on survival, growth and reproduction. Subsequently, full life cycle toxicity assessment work has been extended to other chemicals and fish species.

OECD Test No. 210 (2013a) has been standardized to define the lethal and sublethal effects of chemicals on life cycle stages in fish. The most commonly used test organism is *D. rerio*, but *P. promelas*, *O. latipes* and *O. mykiss* can also be tested. Chronic lethal and sublethal effects are estimated by exposing fish to a range of concentrations of the chemical dissolved in water. Exposure begins immediately after fertilization of the eggs, before the cleavage of the blastodisc begins. During the test, observations and records are made of embryonic development, hatching, survival, abnormal development and behavior.

Using *D. rerio*, *O. mykiss* or *O. latype* as test organisms, the OECD 215 (2000) test describes a method of exposing juvenile fish in the exponential growth phase to a range of sublethal concentrations of the test substance, preferably applied in flow conditions, for 28 days (depending on the stability of the test substance, the frequency of medium renewal varies in semi-static tests, daily renewal being recommended). Fish are examined every 24 during the exposure timeframe and external abnormalities such as hemorrhage or discoloration and abnormal behavior are recorded. Effects on growth rates are analyzed using a regression model to estimate the concentration that would cause x% change in growth rate, ECx (EC10, EC20 or EC30).

A similar test in which the toxic effects of various substances or mixtures on salmonids are tested has been developed by Environment Canada. The test protocol includes three variants: embryo tests for routine monitoring, embryo/fry tests to evaluate the effects on several stages of development and an embryo/fry/fry test for definitive investigations (Environment Canada 2007).

In EU member states, chemicals, municipal and industrial effluents are subject to control for potential toxic effects on fish (Scholz et al. 2013). In Europe, effluent toxicity is evaluated on zebrafish eggs/embryos, the test methodology being internationally available (ISO 2007).

Test Methods 1000.0 and 1001.0 (EPA 2002a, b, c, d, e) for survival and teratogenicity of newly hatched larvae and embryo-larvae of *P. promelas*, respectively, evaluate the chronic toxicity of whole effluents and receiving water in a seven-day static renewal test. The synergistic, antagonistic and additive effects of all chemical, physical and biological components affecting the physiological and biochemical functions of the test organisms are evaluated. The outcomes of this test are total mortality, number of dead embryos and dead and deformed larvae.

The relatively high cost and logistical difficulty of chronic tests on fish with apical objectives (growth and development) limits their application.

A two-generation test developed by EPA in 2015 (MEOGRT, US EPA No. 740-C-15-002) characterizes the effects of a chemical on the endocrine system in Japanese medaka. In both, the F0 and F1 generations, fecundity, fertility and hatching success are evaluated daily, for 21 days. At the end of the test, growth parameters and secondary sex characteristics are recorded, and at subadult life-stage of F1, fish are sampled for growth, secondary sex characteristics, gonad phenotype, and liver vitellogenin (vitellogenin 1; VTG1) mRNA assessment or liver vitellogenin protein assessment.

OECD 240 is a complex test adopted in (2023), similar to EPA 740-C-15-002, performed on specimens of Japanese medaka exposed for several generations (adult fish—parental generation F0, life cycle of the entire future generation F1, continuing until the hatching of the second generations F2) to the action of certain chemicals. The test provides ecologically relevant data on the hazard assessment of some chemicals, including endocrine disruptors chemicals.

Due to the increasing presence of exogenous disruptors that interfere with the endocrine system, OECD has designed tests for screening and testing their negative potential in different fish species (*P. promelas*, *O. latype*, *B. rerio*), respectively OECD Test No. 229 (2012a) and OECD Test No. 230 (2009a). During the in vivo screening (21 days), sexually mature male and female fish are exposed together to a chemical. For all three species, the indicators of endocrine disrupting activity have vitellogenin, and secondary sexual characteristics are biomarkers only for fathead minnow and Japanese medaka). At the end of the test, the gonads are taken for histopathological testing. In the OECD Test No. 230 test (2009a) (21-day Fish Assay. A Short-Term Screening for Oestrogenic and Androgenic Activity, and Aromataze Inhibition), depending on the species used, the measurement of estrogenic, aromatase or androgenic activity of the tested chemical represents biomarker endpoints.

Another test that looks at the effect of endocrine disruptors that interfere with sexual development in fish is the OECD Test No. 234 test (2011b), where fish are exposed to toxic substances during their most sensitive periods. The primary endpoints are developmental aberrations, VTG concentrations and sex ratios (sex ratios) determined by gonadal histology.

Exposure of fish to various chemicals through food, applies to substances with very low solubility in water. In this sense, the OECD Test No. 305 (2012b) test method was developed, which consists of the exposure phase (absorption) to one or more determined concentrations and the post-exposure phase (depuration) of fish species *O. mykiss*, *C. carpio* or *P. promelas*. The tested species are characterized by a distinct feeding behavior (rapid consumption of the administered food ration).

A brief review of freshwater fish toxicity test guidelines published by OECD (the Organization for Economic Co-operation and Development) is presented in Table 4.3.

Since vision is an important sense in zebrafish embryos, and eye defects are easily observed, these are tracked in the US EPA ToxCast test (Sipes et al. 2011). Easier to detect by simple optical techniques are the changes induced in the lateral line system (Froehlicher et al. 2009; Segner 2009).

In toxicological tests, a series of functional and behavioral changes induced by the tested substances can also be followed. Behavioural, blood, respiratory changes (frequency of respiratory movements and oxygen consumption) have been reported

Table 4.3 The different freshwater fish toxicity test guidelines published by OECD

Name of the test	Number of the test	Exposure timeframe	Species	Endpoints
Fish, acute toxicity test	OECD test no. 203	Short-term 96 ore	*Danio rerio* *Pimephales promelas* *Cyprinus carpio* *Oryzias latipes* *Poecilia reticulata* *Lepomis macrochirus* *Oncorhynchus mykiss*	Mortality
Fish, early-life stage toxicity test	OECD test no. 210	30 days post-hatch	*Danio rerio*	Growth, length, hatching, appearance, behavior, mortality
		32 days from start of test (or 28 days post-hatch)	*Pimephales promelas*	
		30 days post-hatch	*Oryzias latipes*	
		2 weeks after controls are free-feeding (or 60 days post-hatch)	*Oncorhynchus mykiss*	
Fish, short-term toxicity test on embryo and sac-fry stages	OECD test no. 212	Short-term test	*Oncorhynchus mykiss* *Brachidanio rerio* *Cyprinus carpio* *Oryzias latipes* *Pimephales promelas* *Carassius auratus* *Lepomis macrochirus*	Hatching, mortality, behavior, appearance
Fish, juvenile growth test	OECD test no. 215	Subchronic	(Oncorhynchus mykiss)	Behavior, appearance, growth
Fish short term reproduction assay	OECD Test No. 229	Subchronic 21 zile	*Pimephales promelas* *Oryzias latipes* *Brachidanio rerio*	Fecundity, yolk protein, sex characteristics
21-day fish assay. A short-term screening for oestrogenic and androgenic activity, and aromatase inhibition	OECD Test No. 230	Subchronic 21 zile	*Pimephales promelas* *Oryzias latipes* *Brachidanio rerio*	Yolk protein, secondary sex characteristics

(continued)

Table 4.3 (continued)

Name of the test	Number of the test	Exposure timeframe	Species	Endpoints
Fish sexual development test	OECD Test No. 234	Subchronic	*Oryzias latipes* *Brachidanio rerio* *Pimephales promelas* *Gasterosteus aculeatus*	Yolk protein, sex ratio
Fish embryo acute toxicity (FET) test	OECD Test No. 236	Acute 96 ore	*Danio rerio*	Mortality, yolk coagulation, heartbeat
Medaka extended one generation reproduction test (MEOGRT)	OECD Test No. 240	Multi-generational 19 weeks	*Oryzias latipes*	Mortality, growth, development, fecundity, yolk protein
Bioaccumulation in fish: aqueous and dietary exposure	OECD Test No. 305	Subchronic 28 days	*Oncorhynchus mykiss* *Cyprinus carpio* *Pimephales promelas*	Bioaccumulation

by numerous authors (El-Sayed and Saad 2008; Paunescu et al. 2022; Ponepal et al. 2023; Velisek et al. 2006; Ullah and Zorriehzahra 2015; Ullah et al. 2019).

Antioxidant enzymes, such as catalase, superoxide dismutase, glutathione S-transferase, malondialdehyde, glutathione peroxidase, glutathione reductase, etc. or their gene expression constitute redox-based biomarkers frequently used in toxicity tests (Alak et al. 2017, 2023; Muthukumaravel et al. 2013; Thenmozhi et al. 2011). Inhibition of brain acetylcholinesterase is also a biomarker often mentioned in studies evaluating the toxicity of some chemical stressors, with reported lethal inhibitions ranging between 40 and 80% (Fergani and Arab 2017; Gibson et al. 1969; Thanomsit et al. 2021; Weiss 1958; Yeltekin 2022).

Cortisol measurement is an excellent method for assessing the stress level of fish (Sadoul and Geffroy 2019), regardless of the research method, i.e., laboratory or field experiments (Geffroy et al. 2018).

The histopathology of various reactive organs is also an aspect followed in many toxicological tests (Teh et al. 1997; Yancheva et al. 2016). Tumors were noticed as a toxicity indicator in in freshwater bullheads (*Amerinons spp.*) from contaminated areas (Fabacher and Baumann 1985). Other authors use Hsp70 (heat shock proteins) and their expression for determining the toxicity of a pollutant (Chen et al. 2022; Jabeen et al. 2022).

According to Walker (2006), biomarkers that indicate a measure of the functioning of a toxic mechanism are called "mechanistic biomarkers", as distinguished from those that only give indications of exposure. The same author appreciates that designing tests that focus on specific biomarkers and critical mechanisms of toxicity for certain types of compounds is preferable to a series of expensive standard toxicity tests (Walker 2006).

Physiological, biochemical and histological measurements have a relatively low applicability as diagnostic tests in aquatic toxicology, generating data that differ from those at the ecosystem level and cannot be unequivocally correlated (Lyubenova and Boteva 2016a, b; Mehrle and Mayer 1980).

Water quality, and implicitly its availability, has always been a cause of concern for human societies (McLennan 2021). Substances that reach the water through different routes can be both cytotoxic and genotoxic, representing factual risks for human health and environment (Ohe et al. 2004). The use of fish as indicators for the cytogenotoxic potential of water, as well as the correlation of results with plant and mammalian test systems, is widely recognized (de Lapuente et al. 2015; Leme and Marin-Morales 2009; Martins and Costa 2017).

Cytogenotoxic effects of various types of waters and chemical agents were evaluated using micronucleus and nuclear abnormalities, and comet assay, together with abnormalities of the cytoplasm. Micronucleus and nuclear abnormalities are considered biomarkers of genotoxicity and genomic instability. It has been suggested that the formation of micronucleus and nuclear abnormalities is due to DNA amplification, sometimes selectively located at the periphery of the nucleus, alterations in the chromatin condensation pattern and architecture, or may be caused by a delayed mitosis that affects a certain amount of DNA, so that the nuclei change their normal oval shape and show protrusions and constrictions (Grønbaek et al. 2007; Fenech et al. 2011). Micronuclei and binuclei frequency in liver cells, gill cells and erythrocytes were recorded to evaluate the effect of municipal wastewater to freshwater fish *Labeo rohita* (Kaur and Dua 2016).

The cytogenotoxic potential of water collected from the urban and rural marshes was assessed by alkaline comet assay using *Oreochromis niloticus* erythrocytes (Pepey et al. 2023). The DMSO-solubilized C_{60} nanomaterial were investigated in *Anabas testudineus*. Nuclear and cytoplasm abnormalities, such as sticky cells, vacuolated cytoplasm, anisochromatic cells and abnormal erythrocyte membrane, along with a significant increase of micronucleus frequency and a significant increase in DNA damage using the comet percent tail DNA were noticed at sublethal concentrations (Sumi and Chitra 2019).

Apoptosis is a type of programmed cell death, a well-controlled vital process, different from necrosis and traumatic cell death, characterized mainly by cytoplasmic condensation, chromatin condensation, DNA fragmentation, membrane blebbing and apoptotic bodies (AnvariFar et al. 2017). DNA damage and apoptosis were noticed in zebrafish embryos as a consequence of the hypoxic conditions triggered by microplastics in waste water (Anifowoshe et al. 2022). Some biomarkers need further investigations and could complete the list of biomarkers in water monitoring programs and treatment strategies.

Cytotoxicity tests performed on in vitro cell cultures are an alternative to acute toxicity tests on fish (Schirmer 2006).

Although approximately 500 fish cell lines are known (Lakra et al. 2011), only several are publicly available through cell banks such as American Tissue Culture Collection (ATCC) or European Collection of Authenticated Cell Cultures (ECACC). Among them an even smaller number of cell lines, such as RTG-2, RTL-W1 and RTgill-W1 from *Oncorhynchus mykiss*, PLHC-1 from *Poeciliopsis lucida*, CHSE-14 from *Oncorhynchus tshawytscha* (Behrens et al. 2001; Bols et al. 2005) was used for cytotoxicity studies. Toxicity of test substances is assessed by cell death caused by the test substance by altering membrane integrity or altering mitochondrial mechanisms (Ekwall 1995; Bols et al. 2005; Vinken and Blaauboer 2017). Fish cell lines have been used to study the modes of action of environmental contaminants, namely DNA damage, alteration of membrane transport, phagocytosis and synthesis of specific proteins, induction of oxidative stress (Bols et al. 2005; Lillicrap et al. 2016).

The rainbow trout gill cell line is used to predict of fish acute toxicity (Tanneberger et al. 2013). For the time being, in order to be accepted as internationally recognized test standards, RTgill-W1 is undergoing formal validation processes within the International Organization for Standardization. The authors demonstrated a very good correlation between the effective concentrations observed in rainbow trout gill cell line and the acute lethal concentration in fish.

The method described by Kawano et al. (2010) with a rainbow trout intestinal epithelial cell line (RTgutGC) can be used to investigate the uptake of toxic substances by the intestinal epithelium. Ex vivo methods can also be used to estimate in vivo absorption: with isolated and perfused fish intestines (Kleinow et al. 1998) and gills (Barron et al. 1989) these tissues require a high degree of technical expertise to isolate and are only viable for a short time.

Fluidic biochips with fish cell lines are being developed for the rapid evaluation of water safety (Brennan et al. 2012).

However, fish cell lines are not yet regulated. Schirmer (2006) recommended selecting tissue-derived cell lines that reflect the site and mode of action of the test chemical to increase the sensitivity of the cellular response by modifying the culture medium (to more closely resemble in vivo exposure) and to account for the fraction of the chemical available to the cells.

Stadnicka-Michalak et al. (2015) showed that tests using a fish cell line, combined with mechanism-based computational models, could replace tests using juvenile fish, the results obtained being well correlated with those generated by in vivo exposure of two fish species to pesticides.

In recent decades, transgenic fish embryo models have also been developed for use in ecotoxicological tests targeting endocrine disorders, such as transgenic medaka strain (ChgH-GFP) harboring the green fluorescence protein (GFP) gene, which enables the activity of the estrogen axis to be visualized in vivo (Spirhanzlova et al. 2016) and Spiggin-GFP medaka fish line was created to identify disruption of the androgen axis (Sebillot et al. 2014).

4.10 Ecotoxicological Tests on Amphibians

Numerous ecotoxicological studies aim to identify the effects of the pollutants on terrestrial and aquatic organisms. Thus, Ferrari et al (2005) proposes several organisms as sentinels in bioassays; among them, larval and adult amphibians shown to be suitable in assays for the evaluation of the toxicity of both xenobiotics and environmental samples (Bantle 1995; Devillers and Exbrayat 1992; Schuytema and Nebeker 1996; Sparling et al. 2000).

Amphibians can constitute bioindicator species of water pollution because embryonic development takes place in water and larval stages present gills (David and Kartheek 2015). Also, adults have a permeable skin, which makes them much more vulnerable in the presence of pollutants than other aquatic organisms (Khan 2021).

The main obstacle in studies evaluating the toxicity of various chemicals on tadpoles is represented by the difficulty of their systematic identification (Rossa-Feres and Nomura 2006). Thus, the first standardized test, which used *Xenopus laevis* embryos as a test organism, was the Frog Embryo Teratogenesis Assay-Xenopus (FETAX) introduced by Dumont in 1983 (ASTM in 1991, 1998; ASTM International 2016). Being an affordable and quick test (lasts 96 h), FETAX is applied as a screening method and aims to evaluate the toxicity of a compound on embryonic development. Through this test, the mortality rate and the type and frequency of malformations are evaluated (Mouche et al. 2011; Xu et al. 2019).

In the last 10 years, the OECD targeted the development and implementation of tests to assess the toxicity of certain chemical substances in amphibians. Thus, the Amphibian Metamorphosis Assay (AMA) is implemented, which uses tadpoles to identify the actions of endocrine disruptors on the normal function of the hypothalamic-pituitary-thyroid (HPT) axis. In this test, stage 51 *Xenopus laevis* tadpoles were subjected, for 21 days, to the action of a chemical substance tested at 3 different concentrations. The observed indicators were: hind limb length, snout to vent length (SVL), development stage, wet weight, thyroid gland histology, mortality (OECD Test No. 231 2009b). Following this test, it is found that substances with prothyroid action stimulate metamorphosis, while antithyroid substances slow down metamorphosis and can induce changes in the structure of the thyroid gland.

Another more complex and informative test is the Amphibian Larval Development Test (LAGDA) which monitors development times, metamorphosis, survival, growth and partial reproductive maturation (OECD Test No. 241 2015). This test is considered to provide suggestive information on the adverse effects induced by the tested substances.

Among the biomarkers used in amphibian toxicological tests, the micronucleus test has been applied for more than three decades in tadpoles, generating an early warning of environmental toxicity (Benvindo-Souza et al. 2020). The first study was carried out by Krauter et al. (1987). Other studies analyze the expression of genes in tadpoles exposed to different pollutants. The measurement of some biomarkers, such as cytochrome p450 (cyp1a), metallothionein (mt) and VTG are important tools in assessing the ecological health of batrachian populations and the ecotoxicological impact of exposure to chemicals on freshwater ecosystems (Carlsson and Tydén 2018).

More recently, tests in which anti-predator activity is assessed have been introduced into ecotoxicological assessments. These tests use special apparatuses in which test tadpoles and predators are kept together, and whose behavior is analyzed (Da Costa Araújo and Malafaia 2020).

4.11 Conclusions

Over time, various toxicity tests and their upgraded versions have been developed and improved, with the aim of identifying hazards, assessing risks, quantifying the toxicity of various chemical agents, but also providing reliable data, which can be extrapolated to different levels of biological organization.

The major standardized aquatic tests were developed by the Organization for Economic Cooperation and Development (OECD) and EPA.

In acute aquatic toxicity assessment, the basic indicator is the amount of toxic substance that induces a biological response in 50% of the test subject population (LD_{50}, LC_{50}, TL_{50}, EC_{50}, IC_{50}). In order to establish the accepted limits of toxicity, sunt utilizați indicatorii de toxicitate cronică (NOEC, LOEC, NOEL, LOEL, MAC).

A wide range of organisms can be used in ecotoxicological freshwater tests (the microalgae: *P. subcapitata, Chlorella vulgaris, Chlamydomonas reinhardtii, Chlorella pyrenoidosa, and Scenedesmus quadricauda, the macroalgae: Cladophora glomerata, Oedogonium cardiacum, Spirogyra* sp., *Chara* sp., *and Zygnema* sp., the Bryophytes: *Fontinalis antipyretica, Platyhypnidium riparoides, Vesicularia dubyana, Scapania undulata, and Fontinalis dalecarlica, the Pteridophytes*: *Azolla pinnata, Salvina molesta, Azolla caroliniana, Salvinia natans*, and *Azolla filiculoides*, the flowering plants: *Lemna minor, Myriophyllum spicatum, Hydrilla verticillata, Lemna gibba and Myriophyllum sibiricum, ciliates*: *Paramecium caudatum, P. primaurelia, P. bursaria, gammarids*: *Gammarus. fasciatus, G. pseudolimnaeus, G. lacustris, crustaceans*: *Daphnia magna, Daphnia pulex, chironomid larvae*: *Chironomus. riparius, C. dilutes, C. yoshimatsui, freshwater mussels, freshwater snails*: *Lymnaea. stagnalis, L. variegates, P. antipodarum, fish-Pimephales promelas, Oryzias latipes, Danio rerio, O. mykiss, O. latype*, amphibians: *Xenopus laevis tadpoles*.

In addition to the mono-species test, recently, multi-species tests are increasingly used, which more accurately reflect the effects and fate (fate) of stressors in modeled ecosystems (macrocosms, mesocosms and microcosms types). Computational modeling models and cytotoxicity tests successfully complete the list of toxicity evaluation tests.

In recent decades, the aim is to reduce the duration of standard toxicological protocols without compromising their predictive value.

Microbiotests are very useful in acute ecotoxicological investigations for the prescreening and screening of chemicals and wastewaters (due to the growing demand for cost-effective tests performed at different levels of biological organization).

Since the appearance of ecotoxicological tests, there has been an increase in the degree of their complexity, from the simplest, in which only the survival of the tested organisms is sought, to the monitoring of some physiological, hematological, biochemical parameters, up to ecological modeling (for example QSAR models, DEBtox model) and the genetic effects in the following generations, in order to achieve a more accurate picture of the impact of toxic substances on aquatic ecosystems. Particular importance is given to testing the first stages in the life of the animals, which are, as a rule, the most sensitive to toxic substances.

The selection of valid test organism species, the identification of new toxicity endpoints in standardized toxicity tests, as well as the development of alternative methods are necessary for an exhaustive investigation of the growing number of chemicals/xenobiotics with potential negative impact on the environment.

Acknowledgements The authors gratefully acknowledge the financial support of the Romanian Ministry of Research, Innovation and Digitization, CNCS/CCCDI-UEFISCDI, through the Project number PN-III-P4-ID-PCE-2020-0620, within PNCDI III.

References

Adams WJ (1995) Aquatic toxicology testing methods. In: Hoffman DJ, Rattner BA, Burton GA Jr, Cairns J Jr (eds) Handbook of ecotoxicology. Lewis Publishers, London, pp 25–46

Adams WJ, Rowland CD (2002) Aquatic toxicology test methods. In: Hoffman DJ, Barnett AR, Burton GA Jr, Cairns J Jr (eds) Handbook of ecotoxicology, 2nd edn. CRC/Lewis Publishers, Boca Raton, FL

AFNOR—Association Française de Normalisation (1996) XP T 90-337: Détermination de l´inhibition de la croissance de *Lemna minor*. https://www.iso.org/obp/ui/#iso:std:iso:20079:ed-1:v1:en. Accessed 24 July 2023

Alak G, Ucar A, Parlak V et al (2017) Assessment of 8-hydroxy-2-deoxyguanosine activity, gene expression and antioxidant enzyme activity on rainbow trout (*Oncorhynchus mykiss*) tissues exposed to biopesticide. Comp Biochem Physiol C Toxicol Pharmacol 203:51–58. https://doi.org/10.1016/j.cbpc.2017.10.007

Alak G, Yeltekin AÇ, Köktürk M et al (2023) May PdCu@f-MWCNT NPs be an ecotoxicologic risk? Appl Organomet Chem 37(3):e7013. https://doi.org/10.1002/aoc.7013

Ali H, Khan E (2018) Bioaccumulation of non-essential hazardous heavy metals and metalloids in freshwater fish. Risk to human health. Environ Chem Lett 16:903–917. https://doi.org/10.1007/s10311-018-0734-7

Alkas FB, Shaban JA, Sukuroglu AA et al (2017) Monitoring and assessment of heavy metal/metalloid concentration by inductively coupled plasma mass spectroscopy (ICP-MS) method in Gonyeli Lake, Cyprus. Environ Monit Assess 189(10):516. https://doi.org/10.1007/s10661-017-6222-x

Allan BM, Nimmo DG, Ierodiaconou D et al (2018) Futurecasting ecological research: the rise of technoecology. Ecosphere 9(5):e02163. https://doi.org/10.1002/ecs2.2163

Alsaadi FM, Madison BN, Brown RS et al (2018) Morphological and molecular effects of two diluted bitumens on developing fathead minnow (*Pimephales promelas*). Aquat Toxicol 204:107–116. https://doi.org/10.1016/j.aquatox.2018.09.003

Amanchi NR (2010) A low cost microbiotest for screening behavioural and ecotoxicological responses of *Paramecium caudatum* and *Oxytricha fallax* to azadirachtin. Adv Appl Sci Res 1(2):124–131

Amaroli A, Trielli F, Sifredi F et al (2010) Nitric oxide production inhibited by xenobiotic compounds in the protozoan *Paramecium primaurelia*. Ecol Indic 10:212–216. https://doi.org/10.1016/j.ecolind.2009.04.012

Anderson BS, Nicely P, Gilbert K et al (2004) Overview of freshwater and marine toxicity tests: a technical tool for ecological risk assessment. California environmental protection agency office of environmental health hazard assessment, Sacramento, CA, pp 64–66, 128–129

Anifowoshe AT, Roy D, Dutta S et al (2022) Evaluation of cytogenotoxic potential and embryotoxicity of KRS-Cauvery river water in zebrafish (*Danio rerio*). Ecotoxicol Environ Saf 233:113320. https://doi.org/10.1016/j.ecoenv.2022.113320

Ankley GT, Villeneuve DL (2006) The fathead minnow in aquatic toxicology: past, present and future. Aquat Toxicol 78(1):91–102. https://doi.org/10.1016/j.aquatox.2006.01.018

AnvariFar H, Amirkolaie AK, Miandare HK et al (2017) Apoptosis in fish: environmental factors and programmed cell death. Cell Tissue Res 368(3):425–439. https://doi.org/10.1007/s00441-016-2548-x

Apostol S (1973) A bioassay of toxicity using protozoa in the study of aquatic environment pollution and its prevention. Envir Res 6(4):365–372. https://doi.org/10.1016/0013-9351(73)90051-0

Arnot JA, Gobas FAPC (2004) A food web bioaccumulation model for organic chemicals in aquatic ecosystems. Environ Toxicol Chem 23(10):2343. https://doi.org/10.1897/03-438

Ashauer R, Boxall AB, Brown CD (2007) New ecotoxicological model to simulate survival of aquatic invertebrates after exposure to fluctuating and sequential pulses of pesticides. Environ Sci Technol 41:1480–1486. https://doi.org/10.1021/es061727b

Association Française de Normalisation (1996) https://www.iso.org/obp/ui/#iso:std:iso:20079:ed-1:v1:en. Accessed 24 July 2023

ASTM (1991) Standard guide for conducting the frog embryo teratogenesis assay—Xenopus (FETAX). ASTM E1439—91. In: Annual book of ASTM standards, Philadelphia

ASTM (1998) Standard guide for conducting the frog embryo teratogenesis assay—Xenopus (FETAX). ASTM E1439—98. In: Annual book of ASTM standards, Philadelphia

ASTM E2455-06 (2013) Standard guide for conducting laboratory toxicity tests with freshwater mussels (Withdrawn 2022). https://www.astm.org/e2455-06r13.html

ASTM International (2016) Standard guide for conducting the frog embryoteratogenesis assay-Xenopus. E-1439. In: Annual book of ASTM standards, vol 11.06. Philadelphia, PA, USA

Bai C, Tang M (2020) Toxicological study of metal and metal oxide nanoparticles in zebrafish. J Appl Toxicol 40:37–63. https://doi.org/10.1002/jat.3910

Bantle JA (1995) FETAX-A developmental toxicity assay using frog embryos. In: Rand GM (ed) Fundamentals of aquatic toxicology. Effects, environmental fate and risk assessment. Taylor and Francis, Washington, pp 207–230

Barber MC (2008) Dietary uptake models used for modeling the bioaccumulation of organic contaminants in fish. Environ Toxicol Chem 27:755–777

Barber MC, Suarez LA, Lassiter RR (1988) Modelling bioconcentration of nonpolar organic pollutants by fish. Environ Toxicol Chem 7:545–558. https://doi.org/10.1002/etc.5620070703

Barber MC, Suarez LA, Lassiter RR (1991) Modelling bioaccumulation of organic pollutants in fish with an application to PCBs in Lake Ontario salmonids. Can J Fish Aquat Sci 48:318–337. https://doi.org/10.1139/f91-044

Barron MG, Schultz IR, Hayton WL (1989) Presystemic branchial metabolism limits di-2-ethylhexulphalate accumulation in fish. Toxicol Appl Pharmacol 98:49–57. https://doi.org/10.1016/0041-008x(89)90133-6

Bastos AC, Prodana M, Abrantes N et al (2014) Potential risk of biochar-amended soil to aquatic systems: an evaluation based on aquatic bioassays. Ecotoxicology 23:1784–1793. https://doi.org/10.1007/s10646-014-1344-1

Behrens A, Schirmer K, Bols NC et al (2001) Polycyclic aromatic hydrocarbons as inducers of cytochrome P4501A enzyme activity in the rainbow trout liver cell line, RTL-W1, and in primary cultures of rainbow trout hepatocytes. Environ Toxicol Chem 20:632–643. https://doi.org/10.1002/etc.5620200324

Benvindo-Souza M, Santos Oliveira EA, Assis RA et al (2020) Micronucleus test in tadpole erythrocytes: trends in studies and new paths. Chemosphere 240:124910. https://doi.org/10.1016/j.chemosphere.2019.124910

Bergquist BL, Bovee EC (1974) A rapid method for monitoring detrimental effects of chemicals on protozoan locomotor rates. J Protozool 21:439

Birge WJ, Black JA, Westerman AG (1985) Short-term fish and amphibian embryo-larval tests for determining the effects of toxicant stress on early life stages and estimating chronic values for single compounds and complex effluents. Environ Toxicol Chem 4:807–821. https://doi.org/10.1002/etc.5620040612

Birge WJ, Black JA. In situ acute/chronic toxicological monitoring of industrial effluents for the NPDES biomonitoring program using fish and amphibian embryo-larval stages as test organisms. Final report. United States: N. p 1981. https://www.osti.gov/biblio/80591. Accessed 28 July 2023

Blaise C, Férard JF (2006) Microbiotests in aquatic toxicology: the way forward, WIT transactions on biomedicine and health. WIT Press, Southampton. https://doi.org/10.2495/ETOX060341

Bodnar IS, Cheban EV (2023) Joint effects of gamma radiation and zinc on duckweed *Lemna minor* L. Aq Toxicol 257:106438. https://doi.org/10.1016/j.aquatox.2023.106438

Bols NC, Dayeh VR, Lee LEJ et al (2005) Use of fish cell lines in the toxicology and ecotoxicology of fish: piscine cell lines in environmental toxicology. In: Mommsen TP, Moon TW (eds) Biochemistry and molecular biology of fishes, vol 6. Elsevier, Amsterdam, pp 43–84

Borgmann U, Grapentine L, Norwood WP et al (2005) Sediment toxicity testing with the freshwater amphipod *Hyalella azteca*: relevance and application. Chemosphere 61(11):1740–1743; author reply 1744–1745. https://doi.org/10.1016/j.chemosphere.2005.04.011

Bour A, Mouchet F, Cadarsi S et al (2016) Toxicity of CeO_2 nanoparticles on a freshwater experimental trophic chain: a study in environmentally relevant conditions through the use of mesocosms. Nanotoxicology 10(2):245–255. https://doi.org/10.3109/17435390.2015.1053422

Braunbeck T, Kais B, Lammer E et al (2015) The fish embryo test (FET): origin, applications, and future. Environ Sci Pollut Res 22:16247–16261. https://doi.org/10.1007/s11356-014-3814-7

Breitholtz M, Rudén C, Ove Hansson S et al (2006) Ten challenges for improved ecotoxicological testing in environmental risk assessment. Ecotoxicol Environ 63:324–335. https://doi.org/10.1016/j.ecoenv.2005.12.009

Brennan LM, Widder MW, Lee LEJ et al (2012) Long-term storage and impedance-based water toxicity testing capabilities of fluidic biochips seeded with RTgill-W1 cells. Toxicol in Vitro 26:736–745. https://doi.org/10.1016/j.tiv.2012.03.010

Bruggeman WA, Matron LBJM, Kooiman D et al (1981) Accumulation and elimination kinetics of di-, tri-, and tetrachlorobiphenyls by goldfish after dietary and aqueous exposure. Chemosphere 10:811–832. https://doi.org/10.1016/0045-6535(81)90082-5

Burton AG Jr, Denton DL, Ho K, Scott Ireland D (2002) Sediment toxicity testing: issues and methods. In: Hoffman DJ, Rattner BA, Burton GA Jr, Cairns J Jr (eds) Handbook of ecotoxicology, 2nd edn. Lewis Publishers, London

Environment Canada (2007) Biological test method: test for measuring the inhibition of growth using the freshwater Macrophyte, *Lemna minor*. EPS 1/RM/37. https://www.canada.ca/content/dam/eccc/migration/main/faunescience-wildlifescience/1ad45620-6a99-470c-8fdc-d57351d47ec8/rm37-202nded-lemnaenglish-20-20u.pdf. Accessed 23 July 2023

Caquet T, Lagadic L, Jonot O et al (1996) Outdoor experimental ponds (mesocosms) designed for long-term ecotoxicological studies in aquatic environment. Ecotoxicol Environ Saf 34:125–133. https://doi.org/10.1006/eesa.1996.0053

Carlsson G, Tydén E (2018) Development and evaluation of gene expression biomarkers for chemical pollution in common frog (*Rana temporaria*) tadpoles. Environ Sci Pollut Res 25:33131–33139. https://doi.org/10.1007/s11356-018-3260-z

Castaño-Sánchez A, Malard F, Kalcíková G et al (2021) Novel protocol for acute in situ ecotoxicity test using native crustaceans applied to groundwater ecosystems. Water 13:1132. https://doi.org/10.3390/w13081132

Ceschin S, Bellini A, Scalici M (2021) Aquatic plants and ecotoxicological assessment in freshwater ecosystems: a review. Environ Sci Pollut Res 28:4975–4988. https://doi.org/10.1007/s11356-020-11496-3

Chapman PM (1990) The sediment quality triad approach to determining pollution-induced degradation. Sci Total Environ 97–98:815–825. https://doi.org/10.1016/0048-9697(90)90277-2

Chen F, Schüürmann G (1997) Quantitative structure-activity relationships in environmental sciences. SETAC Press, Pensacola, FL

Chen W, Zhang M, Luo X et al (2022) Molecular characterization of heat shock protein 20 (hsp20) in goldfish (*Carassius auratus*) and expression analysis in response to environmental stresses. Aquac Rep 24:101106. https://doi.org/10.1016/j.aqrep.2022.101106

Chowdhury KF, Hall RJ, McNally A et al (2023) Phytoremediation as a tool to remove drivers of antimicrobial resistance in the aquatic environment. Rev Environ Contam Toxicol 261:16. https://doi.org/10.1007/s44169-023-00039-9

Clark KE, Gobas FAPC, Mackay D (1990) Model of organic chemical uptake and clearance from fish from food and water. Environ Sci Technol 24:1203–1213. https://doi.org/10.1021/es00078a008

Connolly JP, Pedersen CJ (1988) A thermodynamic-based evaluation of organic chemical accumulation in aquatic organisms. Environ Sci Technol 22:99–103. https://doi.org/10.1021/es00166a011

Cui R, Kwak JI, An Y-J (2023) Multigenerational effects of microplastic fragments derived from polyethylene terephthalate bottles on duckweed *Lemna minor*: Size-dependent effects of microplastics on photosynthesis. Sci Total Environ 872:162159. https://doi.org/10.1016/j.scitotenv.2023.162159

Da Costa Araújo AP, Malafaia G (2020) Can short exposure to polyethylene microplastics change tadpoles' behavior? A study conducted with neotropical tadpole species belonging to order Anura (*Physalaemus cuvieri*). J Hazard Mater 391:122214. https://doi.org/10.1016/j.jhazmat.2020.122214

David M, Kartheek RM (2015) Malathion acute toxicity in tadpoles of *Duttaphrynus melanostictus*, morphological and behavioural study. J Basic Appl Zool 72:1–7. https://doi.org/10.1016/j.jobaz.2015.01.004

de Lapuente J, Lourenço J, Mendo SA et al (2015) The comet assay and its applications in the field of ecotoxicology: a mature tool that continues to expand its perspectives. Front Genet 6:180. https://doi.org/10.3389/fgene.2015.00180

Devillers J, Exbrayat JM (1992) Handbook of ecotoxicological data. Ecotoxicity of chemical to amphibians vol I. Gordon and Breach Science Publisher, Great Britain

DIN 38415-6 (2003) German standard methods for the examination of water, waste water and sludge—subanimal testing (group T)—part 6: toxicity to fish; determination of the non-acute-poisonous effect of waste water to fish eggs by dilution limits (T 6). https://global.ihs.com/doc_detail.cfm?document_name=DIN%2038415%2D6anditem_s_key=00410208

Directive 2010/63/EU of the European Parliament and of the Council of 22 September 2010 on the protection of animals used for scientific purposes. Text with EEA relevance (europa.eu). https://eur-lex.europa.eu/LexUriServ/LexUriServ.do?uri=OJ:L:2010:276:0033:0079:en:PDF

DiToro DM, Zarba CS, Hansen DJ et al (1991) Technical basis for establishing sediment quality criteria for nonionic organic chemicals by using equilibrium partitioning. Environ Toxicol Chem 10:1541–1583. https://doi.org/10.1002/etc.5620101203

Dutka BJ, Jones K, Kwan KK et al (1988) Use of microbial and toxicant screening tests for priority site selection of degraded areas in water bodies. Water Res 22:503–510. https://doi.org/10.1016/0043-1354(88)90047-4

EC [European Commission] (2000) Directive 2000/60/EC of the European parliament and of the council of 23rd October 2000 establishing a framework for community action in the field of water policy official journal of the European communities, L327/1. European Commission, Brussels

ECA (2014) The use of alternatives to testing on animals for the REACH regulation—second report under article 117(3) of the REACH regulation. Helsinki, Finland

ECETOC (2007) Intelligent testing strategies in ecotoxicology: mode of action approach for specifically acting chemicals. European centre for ecotoxicology and toxicology of chemicals technology. Report no. 102, p 182. https://www.ecetoc.org/wp-content/uploads/2021/10/ECETOC-TR-102.pdf. Accessed 26 July 2023

ECHA (2008) Guidance on information requirements and chemical safety assessment chapters R2–R7: information requirements. European chemicals agency guidance for the implementation of REACH. https://echa.europa.eu/guidance-documents/guidance-on-information-requirements-and-chemical-safety-assessment

EEC (1991) Council directive of 15 July 1991 concerning the placing of plant protection products on the market. Office for official publications of the European communities, p 137. https://eur-lex.europa.eu/LexUriServ/LexUriServ.do?uri=CONSLEG:1991L0414:20090101:EN:PDF

Ekner-Grzyb A, Jurga N, Venâncio C, Grzyb T, Grześkowiak BF, Lopes I (2023) Ecotoxicity of non- and PEG-modified lanthanide-doped nanoparticles in aquatic organisms. Aq Toxicol 259:106548. https://doi.org/10.1016/j.aquatox.2023.106548

Ekwall B (1995) The basal cytotoxicity concept. Altern Methods Toxicol 11:721–726

El-Sayed YS, Saad TT (2008) Subacute intoxication of a deltamethrin-based preparation (Butox) 5% EC) in monosex Nile tilapia, *Oreochromis niloticus* L. Basic Clin Pharmacol Toxicol 102(3):293–299. https://doi.org/10.1111/j.1742-7843.2007.00157.x

Environment Canada (1998) Biological test method: toxicity tests using early life stages of salmonid fish (rainbow trout). Ottawa, Ontario. https://www.canada.ca/en/environment-climate-change/services/wildlife-research-landscape-science/biological-test-method-publications/toxicity-rainbow-trout.html

EPA (1996) 712–C–96–122—OPPTS 850.1500 fish life cycle toxicity. https://www.epa.gov/sites/default/files/2015-07/documents/850-1500.pdf

EPA (2002a) Method 1000.0: fathead minnow, *Pimephales promelas*, larval survival and growth; Chronic Toxicity. https://www.epa.gov/sites/default/files/2015-12/documents/method_1000_2002.pdf

EPA (2002b) Method 1001.0: fathead minnow, *Pimephales promelas*, larval survival and teratogenicity test; Chronic Toxicity. https://www.epa.gov/sites/default/files/2015-12/documents/method_1001_2002.pdf

EPA (2002c) Method 1002.0: daphnid, *Ceriodaphnia dubia*, survival and reproduction test; Chronic Toxicity. https://www.epa.gov/sites/default/files/2015-12/documents/method_1002_2002.pdf

EPA (2002d) Short-term methods for estimating the chronic toxicity of effluents and receiving waters to freshwater organisms, 4th edn. https://www.epa.gov/sites/default/files/2015-08/documents/short-term-chronic-freshwater-wet-manual_2002.pdf

EPA (2002e) Test method 2002.0: methods for measuring the acute toxicity of effluents and receiving waters to freshwater and marine organisms; *Ceriodaphnia dubia* acute toxicity tests with effluents and receiving waters. https://www.nemi.gov/methods/method_summary/9319/

EPA (2015) Endocrine disruptor screening program test guidelines 890.2200: medaka extended one generation reproduction test (MEOGRT) [EPA No. 740-C-15-002]. https://www.regulations.gov/document/EPA-HQ-OPPT-2009-0576-0019

EPA (2016a) OCSPP 850.1020: gammarid amphipod acute toxicity test. https://www.regulations.gov/document/EPA-HQ-OPPT-2009-0154-0040

EPA (2016b) OCSPP 850.1300: daphnid chronic toxicity test. https://www.regulations.gov/document/EPA-HQ-OPPT-2009-0154-0034

EPA (United States Environmental Protection Agency) (2023) ECOTOXicology knowledgebase (ECOTOX). https://cfpub.epa.gov/ecotox/. Accessed 19 July 2023

EPA 1000.0 (2002) https://www.epa.gov/sites/default/files/2015-12/documents/method_1001_2002.pdf

Erickson RJ, Mount DR, Highland TL et al (2022) Acute toxicity of major geochemical ions to fathead minnows (*Pimephales Promelas*): part A—observed relationships for individual salts and salt mixtures. Environ Toxicol Chem 41(9):2078–2094. https://doi.org/10.1002/etc.5390

Esterhuizen M, Pflugmacher S (2023) The green liver concept: green liver systems as low-impact systems for bioremediation using aquatic macrophytes. In: Menone ML, Metcalfe C (eds) The Ecotoxicology of aquatic macrophytes. Environmental contamination remediation and management, Springer, Cham, pp 193–214

European Commission (2000) Water Framework Directive. http://ec.europa.eu/environment/water/water-framework/index_en.html

Fabacher DL, Baumann PC (1985) Enlarged livers and hepatic microsomal mixed-function oxidase components in tumor-bearing brown bullheads from a chemically contaminated river. Environ Toxicol Chem 4:703–710. https://doi.org/10.1002/etc.5620040514

Fairchild JF, La Point TW, Zajicek JL et al (1992) Population-, community- and ecosystem-level responses of aquatic mesocosms to pulsed doses of a pyrethroid insecticide. Environ Toxicol Chem 11:115–129. https://doi.org/10.1002/etc.5620110111

Fenech M, Kirsch-Volders M, Natarajan AT et al (2011) Molecular mechanisms of micronucleus, nucleoplasmic bridge and nuclear bud formation in mammalian and human cells. Mutagenesis 26(1):125–132. https://doi.org/10.1093/mutage/geq052

Fergani H, Arab A (2017) Effects of pollution on the activity of acetylcholinesterase in the brain and heart of *Barbus setivimensis* (fish, Cyprinidae) in El Harrach stream (North of Algeria). Revue d'Écologie 72(1):73–82. https://hal.science/hal-03532754/document

Ferrari L, de la Torre FR, Demichelis SO et al (2005) Ecotoxicological assessment for receiving waters with the premetamorphic tadpoles acute assay. Chemosphere 59:567–575. https://doi.org/10.1016/j.chemosphere.2005.01.045

Fochtman P (2000) Acute toxicity of nine pesticides as determined with conventional assays and alternative microbiotests. In: Persoone G, Janssen C, De Coen W (eds) New microbiotests for routine toxicity screening and biomonitoring. Kluwer Academic, Plenum Publishers, Dordrecht, pp 233–241

Froehlicher M, Liedtke A, Groth KJ et al (2009) Zebrafish (*Danio rerio*) neuromast: promising biological endpoint linking developmental and toxicological studies. Aquat Toxicol 95:307–319. https://doi.org/10.1016/j.aquatox.2009.04.007

Gagnaire B, Boyer P, Bonzom JM et al (2011) Transfer modelling and toxicity evaluation of the effluent from an installation of cleansing and uranium recovery using a battery of bioassays. Ecotoxicology 20(1):187–201. https://doi.org/10.1007/s10646-010-0570-4

Gavrilescu E (2008) General notions of ecotoxicology. Ed. Sitech, Craiova (In Romanian)

Geffroy B, Sadoul B, Bouchareb A et al (2018) Nature-based tourism elicits a phenotypic shift in the coping abilities of fish. Front Physiol 9:13. https://doi.org/10.3389/fphys.2018.00013

Gibson RF, Ludke JL, Ferguson DE (1969) Sources of error in the use of fish-brain acetylcholinesterase as a monitor for pollution. Bull Environ Contam Toxicol 4:17–23. https://doi.org/10.1007/BF01579152

Gilron GL, Lynn DH (1998) Ciliated protozoa as test organisms in toxicity assessments. In: Wells PG, Lee K, Blaise C (eds) Microscale testing in aquatic toxicology. Advances, Techniques and Practice, CRC Publishers, pp 323–336

Girling AE, Pascoe D, Janssen CR et al (2000) Development of methods for evaluating toxicity to freshwater ecosystems. Ecotoxicol Environ Saf 45(2):148–176. https://doi.org/10.1006/eesa.1999.1847

Gobas FAPC (1993) A model for predicting the bioaccumulation of hydrophobic organic chemicals in aquatic food webs: application to Lake Ontario. Ecol Model 69:1. https://doi.org/10.1016/0304-3800(93)90045-T

Grønbaek K, Hother C, Jones PA (2007) Epigenetic changes in cancer. APMIS 115:1039–1059

Hamelink JL, Waybrant RC, Ball RC (1971) A proposal: exchange equilibria control the degree chlorinated hydrocarbons are biologically magnified in lentic environments. Trans Am Fish Soc 100:207–214. https://doi.org/10.1577/1548-8659

Heine S, Schmitt W, Görlitz G et al (2014) Effects of light and temperature fluctuations on the growth of Myriophyllum spicatum in toxicity tests–a model-based analysis. Environ Sci Pollut Res 21:9644–9654. https://doi.org/10.1007/s11356-014-2886-8

Heise S, Babut M, Casado C et al (2020) Ecotoxicological testing of sediments and dredged material: an overlooked opportunity? J Soils Sediments 20:4218–4228. https://doi.org/10.1007/s11368-020-02798-7

Hernández-García CI, Martínez-Jerónimo F (2023) Changes in the morphology and cell ultrastructure of a microalgal community exposed to a commercial glyphosate formulation and a toxigenic cyanobacterium. Front Microbiol 14:1195776. https://doi.org/10.3389/fmicb.2023.1195776

Hill AJ, Teraoka H, Heideman W et al (2005) Zebrafish as a model vertebrate for investigating chemical toxicity. Toxicol Sci 86:6–19. https://doi.org/10.1093/toxsci/kfi110

Hinton DE, Kullman SW, Hardman RC et al (2005) Resolving mechanisms of toxicity while pursuing ecotoxicological relevance? Mar Pollut Bull 51:635–648. https://doi.org/10.1016/j.marpolbul.2005.07.020

Hsu CH, Wen ZH, Lin CS et al (2007) The Zebrafish model: use in studying cellular mechanisms for a spectrum of clinical disease entities. Curr Neurovasc Res 4:111–120. https://doi.org/10.2174/156720207780637234

ISO DIS 20079 (2019) https://www.iso.org/standard/34074.html. Accessed 24 July 2023

ISO (1996a) 7346-1, Water quality—determination of the acute lethal toxicity of substances to a freshwater fish [*Brachydanio rerio* Hamilton-Buchanan (Teleostei, Cyprinidae)]—part 1: static method. https://www.iso.org/obp/ui/#iso:std:iso:12890:ed-1:v1:en2

ISO (1996b) 7346-2, water quality—determination of the acute lethal toxicity of substances to a freshwater fish [*Brachydanio rerio* Hamilton-Buchanan (Teleostei, Cyprinidae)]—part 2: semi-static method. https://www.iso.org/obp/ui/#iso:std:iso:12890:ed-1:v1:en2

ISO (1996c) 7346-3, water quality—determination of the acute lethal toxicity of substances to a freshwater fish [*Brachydanio rerio* Hamilton-Buchanan (Teleostei, Cyprinidae)]—part 3: flow-through method. https://www.iso.org/obp/ui/#iso:std:iso:12890:ed-1:v1:en2

ISO (2007) Water quality—determination of the acute toxicity of wastewater to zebrafish eggs (Danio rerio). ISO 15088:2007(E). Geneva, Switzerland

ISO (2008) 20666: water quality—determination of the chronic toxicity to *Brachionus calyciflorus* in 48 h. https://www.iso.org/standard/39674.html

ISO (2011) 14380: water quality—determination of the acute toxicity to *Thamnocephalus platyurus* (Crustacea, Anostraca). ISO, Geneva, Switzerland. https://www.iso.org/standard/54613.html

ISO (2012) 6341: water quality–determination of the inhibition of the mobility of *Daphnia magna* Straus (Cladocera, Crustacea)—acute toxicity test. ISO: Geneva, Switzerland. https://www.iso.org/standard/54614.html

ISO (2013) 16303: determination of toxicity of fresh water sediments using *Hyalella Azteca*. https://www.iso.org/obp/ui/#iso:std:iso:16303:ed-1:v1:en

Jabeen G, Ishaq S, Arshad M et al (2022) Investigation on immune-related protein (Heat shock proteins and metallothionein) Gene expression changes and liver histopathology in cadmium-stressed fish. Biomed Res Int. https://doi.org/10.1155/2022/2075791

Jabiol J, McKie BG, Bruder A, Bernadet C, Gessner MO, Chauvet E (2013) Trophic complexity enhances ecosystem functioning in an aquatic detritus-based model system. J Anim Ecol 82:1042–1051. https://doi.org/10.1111/1365-2656.12079

Jager T (2020) Revisiting simplified DEBtox models for analysing ecotoxicity data. Ecol Model 416:108904. https://doi.org/10.1016/j.ecolmodel.2019.108904

Jamil Emon F, Rohani MF, Sumaiya N et al (2023) Bioaccumulation and bioremediation of heavy metals in fishes—a review. Toxics 11:510. https://doi.org/10.3390/toxics11060510

Janssen CR, Vangheluwe M, Van Sprang P (2000) A brief review and critical evaluation of the status of microbiotests. In: Persoone G, Janssen C, De Coen W (eds) New microbiotests for routine toxicity screening and biomonitoring. Kluwer Academia/Plenum Publishers, New York, pp 27–37

Kaur R, Dua A (2016) Fish liver and gill cells as cytogenotoxic indicators in assessment of water quality. Environ Sci Pollut Res 23:18892–18900. https://doi.org/10.1007/s11356-016-7080-8

Kawano A, Haiduk C, Schirmer K et al (2010) Development of a rainbow trout intestinal epithelial cell line and its response to lipopolysaccharide. Aquac Nutr. https://doi.org/10.1111/j.1365-2095.2010.00757.x

Kennedy JH, LaPoint TW, Balci P et al (2002) Model aquatic ecosystems in ecotoxicological research: considerations of design, implementation, and analysis. In: Hoffman DJ, Rattner BA, Burton GA Jr, Cairns J Jr (eds) Handbook of ecotoxicology, 2nd edn. Lewis Publishers, London

Khan K (2021) Chemometric modeling of toxicity of chemicals to tadpoles. In: Roy K (ed) Chemometrics and cheminformatics in aquatic toxicology, pp 331–357

Kleinow KM, James MO, Tong Z et al (1998) Bioavailability and biotransformation of benzo[a]pyrene in an isolated perfused in situ catfish intestinal preparation. Environ Health Perspect 106:155–166. https://doi.org/10.1289/ehp.98106155

Komala Z (1995) Notes on the use of invertebrates, especially ciliates, in studies on pollution and toxicity. Folia Biol (krakow) 43(1–3):25–27

Kooijman SALM (1987) A safety factor for LC50 values allowing for differences in sensitivity among species. Water Res 21:269–276. https://doi.org/10.1016/0043-1354(87)90205-3

Krauter PW, Anderson SL, Harrison FL (1987) Radiation-induced micronuclei in peripheral erythrocytes of *Rana catesbeiana*: an aquatic animal model for in vivo genotoxicity studies. Environ Mol Mutagen 10(3):285–296. https://doi.org/10.1002/em.2850100307

Lakra W, Swaminathan TR, Joy K (2011) Development, characterization, conservation and storage of fsh cell lines: a review. Fish Physiol Biochem 37:1–20. https://doi.org/10.1007/s10695-010-9411-x

Landrum PF, Fontaine TD, Faust WR et al (1992) Modeling the accumulation of polycyclic aromatic hydrocarbons by the amphipod *Diporeia spp.* In: Gobas FAPC, McCorquodale JA (eds) Chemical dynamics in fresh water ecosystems. Lewis, Boca Raton, FL, pp 111–128

Law FCP, Abedini S, Kennedy CJ (1991) A biologically based toxicokinetic model for pyrene in rainbow trout. Toxicol Appl Pharmacol 110:390–402. https://doi.org/10.1016/0041-008x(91)90041-c

Leme DM, Marin-Morales MA (2009) *Allium cepa* test in environmental monitoring: a review on its application. Mutat Res Rev Mutat Res 682(1):71–81. https://doi.org/10.1016/j.mrrev.2009.06.002

Lescord GL, Kidd KA, De Silva AO, Williamson M, Spencer C, Wang X, Muir DC (2015) Perfluorinated and polyfluorinated compounds in lake food webs from the Canadian high Arctic. Environ Sci Technol 49:2694–2702. https://doi.org/10.1021/es5048649

Lewis M, Thursby G (2018) Aquatic plants: test species sensitivity and minimum data requirement evaluations for chemical risk assessments and aquatic life criteria development for the USA. Environ Pollut 238:270–280. https://doi.org/10.1016/j.envpol.2018.03.003

Li ZH, Li P, Shi ZC (2015) Responses of the hepatic glutathione antioxidant defense system and related gene expression in juvenile common carp after chronic treatment with tributyltin. Ecotoxicology 24:700–705. https://doi.org/10.1007/s10646-014-1416-2

Li JJ, Yue YX, Jiang JF et al (2023) Assessment of toxic mechanisms and mode of action to three different levels of species for 14 antibiotics based on interspecies correlation, excess toxicity, and QSAR. Chemosphere 317:137795. https://doi.org/10.1016/j.chemosphere.2023.137795

Lillicrap A, Belanger S, Burden N et al (2016) Alternative approaches to vertebrate ecotoxicity tests in the 21st century: a review of developments over the last 2 decades and current status. Environ Toxicol Chem 35(11):2637–2646. https://doi.org/10.1002/etc.3603

Lyubenova M, Boteva S (2016a) Biotests in ecotoxicology: current practice and problems. In: Tech. https://doi.org/10.5772/64776

Lyubenova M, Boteva S (2016). Biotests in ecotoxicology: current practice and problems. Toxicology—new aspects to this scientific conundrum. https://www.intechopen.com/chapters/51891. https://doi.org/10.5772/64776
Lyubenova M, Kalchev R (2009) Ecotoxicology—small practicum (2nd amended edn). An-Di Press, Sofia, p 305
Mahaye N, Musee N (2023) Evaluation of apical and molecular effects of algae *Pseudokirchneriella subcapitata* to cerium oxide nanoparticles. Toxics 11:283. https://doi.org/10.3390/toxics11030283
Martins M, Costa PM (2017) The comet assay in aquatic (eco)genotoxicology using non-conventional model organisms: relevance, constraints and prospects. In: Larramendy ML (ed) Ecotoxicology and genotoxicology: non-traditional aquatic models. Royal Society of Chemistry, pp 1–32
McLennan M (2021) The global risks report 2021, 16th edn. World Economic Forum
Mehrle PM, Mayer FL (1980) Clinical tests in aquatic toxicology: state of the art. Environ Health Perspect 34:139–143. https://doi.org/10.1289/ehp.8034139
Mendieta Herrera J, Iñiguez Armijos C, Rosado Alcarria D et al (2023) Toxicity of difenoconazole and atrazine and their photodegradation products on aquatic biota: environmental implications in countries lacking good agricultural practices. Toxics 11(3):213. https://doi.org/10.3390/toxics11030213
Mills WL (1976) Water quality bioassay using selected protozoa. I. [*Paramecium candatum*; *Amoeba proteus*; *Euglena gracilis*]. United States. https://doi.org/10.1080/10934527609385790. Accessed 2 July 2023
Mitchell ED, Burgess JE, Stuetz RM (2002) Developments in ecotoxicity testing. Rev Environ Sci Biotechnol 1:169–198. https://doi.org/10.1023/A:1020842718996
Miyoshi N, Kawano T, Tanaka M et al (2003) Use of *Paramecium* species in bioassays for environmental risk management: determination of IC50 values for water pollutants. J Health Sci 49(6):429–435. http://jhs.pharm.or.jp/data/49(6)/49_429.pdf
Mohr S, Meinecke S, Feibicke M et al (2023) Effects of a realistic pesticide spraying sequence for apple crop on stream communities in mesocosms: negligible or notable? Environ Sci Eur 35:35. https://doi.org/10.1186/s12302-023-00739-y
Morrison HA, Gobas FAPC, Lazar R et al (1996) Development and verification of a bioaccumulation model for organic contaminants in benthic invertebrates. Environ Sci Technol 30:3377–3384. https://doi.org/10.1021/es960280b
Mouche I, Malesic L, Gillardeaux O (2011) FETAX assay for evaluation of developmental toxicity. Methods Mol Biol 691:257–269. https://doi.org/10.1007/978-1-60761-849-2_15
Mount DI, Stephan CE (1967) A method for establishing acceptable toxicant limits for fish—malathion and 2,4-D. Trans Am Fish Soc 96:185–193. https://doi.org/10.1577/1548-8659
Muir DCG, Yarechewski AL (1988) Dietary accumulation of four chlorinated dioxin congeners by rainbow trout and fathead minnows. Environ Toxicol Chem 7:227–236. https://doi.org/10.1002/etc.5620070306
Muthukumaravel K, Sivakumar B, Kumarasamy P et al (2013) Studies on the toxicity of pesticide monocrotophos on the biochemical constituents of the freshwater fish *Labeo rohita*. Int J Curr Biochem Biotechnol 2:20–26
Nabholz JV, Miller P, Zeeman M (1993) Environmental risk assessment of new chemicals under the toxic substances control act (TSCA) section five. In: Landis WG, Hughes JS, Lewis MA (eds) Environmental toxicology and risk assessment. ASTM Special Technical Publication 1179. ASTM International, Philadelphia, pp 40–55
Nair HT, Perumal S (2002) Trophic transfer and accumulation of microplastics in freshwater ecosystem: risk to food security and human health. Int J Ecol 2022:1234078. https://doi.org/10.1155/2022/1234078
Netzeva TI, Pavan M, Worth AP (2008) Review of (quantitative) structure-activity relationships for acute aquatic toxicity. QSAR Comb Sci 27:77–90. https://doi.org/10.1002/qsar.200710099

Nichols JW, McKim JM, Andersen ME et al (1990) A physiology based toxicokinetic model for the uptake and disposition of waterborne organic chemicals in fish. Toxicol Appl Pharmacol 106:433–447. https://doi.org/10.1016/0041-008x(90)90338-u

Nicolau A, Manuel M, Nelson L (1999) Physiological responses of Tetrahymena pyriformis to copper, zinc, cycloheximide and Triton X100. FEMS Microbiol Ecol 30(3):209–216. https://doi.org/10.1111/j.1574-6941.1999.tb00649.x

Niimi AJ, Oliver BG (1987) Influence of molecular weight and molecular volume on the dietary absorption efficiency of chemicals by fish. Can J Fish Aquat Sci 45:222–227. https://doi.org/10.1139/f88-026

Nisbet RM, Muller EB, Lika K et al (2000) From molecules to ecosystems through dynamic energy budget models. J Anim Ecol 69:913–926. https://doi.org/10.1111/j.1365-2656.2000.00448.x

Norberg TJ, Mount DI (1985) A new fathead minnow (*Pimephales promelas*) subchronic toxicity test. Environ Toxicol Chem 4(5):711–718. https://doi.org/10.1002/etc.5620040515

Norberg-King TJ (1989) An evaluation of the fathead minnow seven-day subchronic test for estimating chronic toxicity. Environ Toxicol Chem 8:1075–1089. https://doi.org/10.1002/etc.5620081113

Odum EP (1984) The mesocosm. BioScience 34(9):558–562. https://www.jstor.org/stable/i256062

OECD (2012c) Test no. 211: daphnia magna reproduction test, OECD guidelines for the testing of chemicals, section 2. OECD Publishing, Paris. https://doi.org/10.1787/9789264185203-en

OECD (1998) Test no. 212: fish, short-term toxicity test on embryo and sac-fry stages, OECD guidelines for the testing of chemicals, section 2. OECD Publishing, Paris. https://doi.org/10.1787/9789264070141-en

OECD (2000) Test no. 215: fish, juvenile growth test, OECD guidelines for the testing of chemicals, section 2. OECD Publishing, Paris. https://doi.org/10.1787/9789264070202-en

OECD (2004) Test no. 202: *Daphnia sp.* acute immobilisation test, OECD guidelines for the testing of chemicals, section 2. OECD Publishing, Paris. https://doi.org/10.1787/9789264069947-en

OECD (2006) Test no. 221: *Lemna sp.* growth inhibition test, OECD guidelines for the testing of chemicals, section 2. OECD Publishing, Paris. https://doi.org/10.1787/9789264016194-en

OECD (2007) Test no. 225: sediment-water lumbriculus toxicity test using spiked sediment, OECD guidelines for the testing of chemicals, section 2. OECD Publishing, Paris. https://doi.org/10.1787/9789264067356-en

OECD (2009a) Test no. 230: 21-day fish assay: a short-term screening for oestrogenic and androgenic activity, and aromatase inhibition. OECD guidelines for the testing of chemicals, section 2. OECD Publishing, Paris. https://doi.org/10.1787/9789264076228-en

OECD (2009b) Test no. 231: amphibian metamorphosis assay, OECD guidelines for the testing of chemicals, section 2. OECD Publishing, Paris. https://doi.org/10.1787/9789264076242-en

OECD (2010) Test no. 233: sediment-water chironomid life-cycle toxicity test using spiked water or spiked sediment, OECD guidelines for the testing of chemicals, section 2. OECD Publishing, Paris. https://doi.org/10.1787/9789264090910-en

OECD (2011a) Test no. 201: freshwater alga and cyanobacteria, growth inhibition test, OECD guidelines for the testing of chemicals, section 2. OECD Publishing, Paris. https://doi.org/10.1787/9789264069923-en

OECD (2011b) Test no. 234: fish sexual development test, OECD guidelines for the testing of chemicals, section 2. OECD Publishing, Paris. https://doi.org/10.1787/9789264122369-en

OECD (2011c) Test no. 235: *Chironomus sp.*, acute immobilisation test, OECD guidelines for the testing of chemicals, section 2. OECD Publishing, Paris. https://doi.org/10.1787/9789264122383-en

OECD (2012a) Test no. 229: fish short term reproduction assay, OECD guidelines for the testing of chemicals, section 2. OECD Publishing, Paris. https://doi.org/10.1787/9789264185265-en

OECD (2012b) Test no. 305: bioaccumulation in fish: aqueous and dietary exposure, OECD guidelines for the testing of chemicals, section 3. OECD Publishing, Paris. https://doi.org/10.1787/9789264185296-en

OECD (2013a) Test no. 210: fish, early-life stage toxicity test, OECD guidelines for the testing of chemicals, section 2. OECD Publishing, Paris. https://doi.org/10.1787/9789264203785-en
OECD (2013b) Test no. 236: fish embryo acute toxicity (FET) test, OECD guidelines for the testing of chemicals, section 2. OECD Publishing, Paris. https://doi.org/10.1787/9789264203709-en
OECD (2014a) Test no. 239: water-sediment myriophyllum spicatum toxicity test, guidelines for the testing of chemicals, section 2. OECD Publishing, Paris. https://doi.org/10.1787/9789264224155-en
OECD (2014b) Test no. 238: sediment-free myriophyllum spicatum toxicity test, OECD guidelines for the testing of chemicals, section 2. OECD Publishing, Paris. https://doi.org/10.1787/9789264224131-en
OECD (2015) Test no. 241: the larval amphibian growth and development assay (LAGDA), OECD guidelines for the testing of chemicals, section 2, OECD Publishing, Paris. https://doi.org/10.1787/9789264242340-en
OECD (2016a) Test no. 242: potamopyrgus antipodarum reproduction test, OECD guidelines for the testing of chemicals, section 2. OECD Publishing, Paris. https://doi.org/10.1787/9789264264311-en
OECD (2016b) Test no. 243: lymnaea stagnalis reproduction test, OECD guidelines for the testing of chemicals, section 2. OECD Publishing, Paris. https://doi.org/10.1787/9789264264335-en
OECD (2018) Chironomid toxicity test using spiked sediment (OECD TG 218) or spiked water (OECD TG 219). In: Revised guidance document 150 on standardised test guidelines for evaluating chemicals for endocrine disruption. OECD Publishing, Paris. https://doi.org/10.1787/9789264304741-11-en
OECD (2019) Test no. 203: fish, acute toxicity test, OECD guidelines for the testing of chemicals, section 2. OECD Publishing, Paris. https://doi.org/10.1787/9789264069961-en
OECD (2023) Test no. 240: medaka extended one generation reproduction test (MEOGRT), OECD guidelines for the testing of chemicals, section 2. OECD Publishing, Paris. https://doi.org/10.1787/9789264242258-en
Ohe T, Watanabe T, Wakabayashi K (2004) Mutagens in surface waters: a review. Mutat Res Rev Mutat Res 567(2–3):109–149. https://doi.org/10.1016/j.mrrev.2004.08.003
Ohmura T, Nishigami Y, Taniguchi A et al (2018) Simple mechanosense and response of cilia motion reveal the intrinsic habits of ciliates. Proc Natl Acad Sci USA 115(13):3231–3236. https://doi.org/10.1073/pnas.1718294115
Onishi Y, Tatarazako N, Koshio T et al (2021) Summary of reference chemicals evaluated by the fish short-term reproduction assay, OECD TG229, using Japanese Medaka, *Oryzias latipes*. J Appl Toxicol 41(8):1200–1221. https://doi.org/10.1002/jat.4104
Oroş V (2002) Biomonitoring. In: Environmental pollution and monitoring, Ed.Univ. Transilvania, Brasov, pp 60–73. (In Romanian)
Padilla S, Cowden J, Hinton DE et al (2009) Use of medaka in toxicity testing. Curr Protoc Toxicol Chapter 1(Unit1):10. https://doi.org/10.1002/0471140856.tx0110s39
Paunescu A, Ponepal CM, Tofan L et al (2022) Ecotoxicological risk assessment of actellic 50 EC insecticide on non-target organisms in parallel with the application of standardized tests. Toxics 10:745. https://doi.org/10.3390/toxics10120745
Pepey E, Conéjéro G, Ndrianaivo EN et al (2023) Environ Toxicol Chem 42(6):1266–1275. https://doi.org/10.1002/etc.5605
Persoone G, Dive D (1978) Toxicity tests on ciliates-a short review. Ecotoxicol Environ Saf 2(2):105–114. https://doi.org/10.1016/0147-6513(78)90001-5
Persoone G, Baudo R, Cotman M et al (2009) Review on the acute daphnia magna toxicity test—evaluation of the sensitivity and the precision of assays performed with organisms from laboratory cultures or hatched from dormant eggs. Knowl Manag Aquat Ecosyst 393:01. https://doi.org/10.1051/kmae/2009012
Petre M, Teodorescu A (2009) Biotechnology of environmental protection. Course notes, 2nd edn. Revised and added. Editura CD Press, pp 113–130 (In Romanian)

Poikane S, Herrero FS, Kelly MG et al (2020) European aquatic ecological assessment methods: a critical review of their sensitivity to key pressures. Sci Total Environ 740:140075. https://doi.org/10.1016/j.scitotenv.2020.140075

Ponepal CM, Soare LC, Drăghiceanu OA et al (2023) Evaluation of the morphological, physiological and biochemical effects induced by coragen 20 SC in some non-target species. Toxics 11:618. https://doi.org/10.3390/toxics11070618

Prabhu J (2017) Frugal innovation: doing more with less for more. Philos Trans Royal Soc A Math Phys Eng Sci 375:20160372. https://doi.org/10.1098/rsta.2016.0372

Rand GM (2008) Fish toxicity studies. In: Di Giulio RT, Hinton DE (eds) The toxicology of fishes, 1st edn. CRC Press, Taylor and Francis Group, Boca Raton, Florida, pp 659–681

Rao JV, Srikanth K, Arepalli SK et al (2006) Toxic effects of acephate on *Paramecium caudatum* with special emphasis on morphology, behaviour, and generation time. Pestic Biochem Physiol 86(3):131–137. https://doi.org/10.1016/j.pestbp.2006.02.005

Rao JV, Gunda VG, Srikanth K et al (2007) Acute toxicity bioassay using *Paramecium caudatum*, a key member to study the effects of monocrotophos on swimming behaviour, morphology and reproduction. Toxicol Environ Chem 89(2):307–317. https://doi.org/10.1080/02772240601010071

Rodgers JH Jr, Crossland NO, Kline ER et al (1996) Design and construction of model stream ecosystems. Ecotoxicol Environ Saf 33:30–37. https://doi.org/10.1006/eesa.1996.0003

Rosenberg DM, Resh VH (1993) Introduction to freshwater biomonitoring and benthic macroinvertebrates. In: Rosenberg DM, Resh VH (eds) Freshwater biomonitoring and benthic macroinvertebrates. Chapman/Hall, New York, pp 1–9

Rossa-Feres DDC, Nomura F (2006) Characterization and taxonomic key for tadpoles (Amphibia: Anura) from the northwestern region of São Paulo State, Brazil. Biota Neotrop 6(1):BN00706012006. https://doi.org/10.1590/S1676-06032006000100014

Roubeau Dumont E, Elger A, Azéma C et al (2022) Cutting-edge spectroscopy techniques highlight toxicity mechanisms of copper oxide nanoparticles in the aquatic plant *Myriophyllum spicatum*. Sci Total Environ 803:150001. https://doi.org/10.1016/j.scitotenv.2021.150001

Roubeau Dumont E, Gao X, Zheng J et al (2023) Unraveling the toxicity of tire wear contamination in three freshwater species: from chemical mixture to nanoparticles. J Hazard Mater 453:131402. https://doi.org/10.1016/j.jhazmat.2023.131402

Roveri V, Guimarães LL, Correia AT (2023) Computational modeling (in silico) methods combined with ecotoxicological experiments (in vivo) to predict the environmental risks of an antihistamine drug (loratadine). Drug Chem Toxicol https://doi.org/10.1080/01480545.2023.2232563

Ruck JG, Martin M, Mabon M (2000) Evaluation of toxkits as methods for monitoring water quality in New Zealand. In: Persoone G, Janssen C, De Coen W (eds) New microbiotests for routine toxicity screening and biomonitoring. Kluwer Academic, Plenum Publishers, Dordrecht, pp 103–119

Saari T, Villeneuve D, Ankley G et al (2016) Bringing the fathead minnow into the genomic era. Midwest SETAC, Madison, WI, March 14–16

Sadoul B, Geffroy B (2019) Measuring cortisol, the major stress hormone in fishes. J Fish Biol 94(4):540–555. https://doi.org/10.1111/jfb.13904

Sánchez-Bayo F, Tennekes HA, Goka K (2013) Impact of systemic insecticides on organisms and ecosystems. In: Trdan S (ed) Insecticides—development of safer and more effective technologies. InTech. https://doi.org/10.5772/52831

Schirmer K (2006) Proposal to improve vertebrate cell cultures to establish them as substitutes for the regulatory testing of chemicals and effluents using fish. Toxicology 224:163–183. https://doi.org/10.1016/j.tox.2006.04.042

Scholz S, Sela E, Blaha L et al (2013) A European perspective on alternatives to animal testing for environmental hazard identification and risk assessment. Regul Toxicol Parmacol 67:506–530. https://doi.org/10.1016/j.yrtph.2013.10.003

Schriever CA, Von der Ohe PC, Liess M (2007) Estimating pesticide runoff in small streams. Chemosphere 68:2161–2171. https://doi.org/10.1016/j.chemosphere.2007.01.086

Schuytema GS, Nebeker AV (1996) Amphibian's toxicity data for water quality criteria chemical. EPA. National Health and Environmental Effects Research Laboratory, Western Ecology Division, Corvallis, Oregon. EPA/600/R/124. US

Sebillot A, Damdimopoulou P, Ogino Y et al (2014) Rapid fluorescent detection of (anti)androgens with spiggin-gfp medaka. Environ Sci Technol 48:10919–10928. https://doi.org/10.1021/es5030977

Segner H (2009) Zebrafish (*Danio rerio*) as a model organism for investigating endocrine disruption. Comp Biochem Phys 149C:187–195. https://doi.org/10.1016/j.cbpc.2008.10.099

Sesin V, Arnolds JL, Arts GHP (2023) Global perspective for the use of aquatic macrophytes in regulatory risk assessment for contaminants. In: Menone ML, Metcalfe C (eds) The ecotoxicology of aquatic macrophytes. Environmental contamination remediation and management. Springer, Cham, pp 75–108

Shunmugam AP, Subramanian G, Fernandez JG (2021) Measurements of the swimming speeds of motile microorganisms using object tracking and their correlation with water pollution and rheology levels. Sci Rep 11:11821. https://doi.org/10.1038/s41598-021-91134-1

Sipes NS, Padilla S, Knudsen TB (2011) Zebrafish: as an integrative model for twenty-first century toxicity testing. Birth Defects Res 93C:256–267. https://doi.org/10.1002/bdrc.20214

Skou PB, Khakimov B, Hansen TH et al (2018) Chemical characterization by gas chromatography-mass spectrometry and inductively coupled plasma-optical emission spectroscopy of membrane permeates from an industrial dairy ingredient production used as process water. J Dairy Sci 101(1):135–146. https://doi.org/10.3168/jds.2017-12950

Smart RM, Barko JW (1985) Laboratory culture of submersed freshwater macrophytes on natural sediments. Aq Bot 21(3):251–263

Sonneborn TM (1970) Methods in *Paramecium* research. In: Prescott DM (ed) Methods in cell physiology. Academic Press, New York, pp 241–339

Sourisseau S, Basseres A, Perie F et al (2008) Calibration, validation and sensitivity analysis of an ecosystem model applied to artificial streams. Water Res 42:1167–1181. https://doi.org/10.1016/j.watres.2007.08.039

Sparling DW, Bishop CA, Linder G (2000) The current status of amphibian and reptile ecotoxicological research. In: Sparling DW, Linder G, Bishop CA (eds) Ecotoxicology of amphibians and reptiles. SETAC Press, Pensacola, pp 1–13

Spirhanzlova P, Leleu M, Sébillot A et al (2016) Oestrogen reporter transgenic medaka for non-invasive evaluation of aromatase activity. Comp Biochem Physiol C Toxicol Pharmacol 179:64–71. https://doi.org/10.1016/j.cbpc.2015.08.011

Stadnicka-Michalak J, Schirmer K, Ashauer R (2015) Toxicology across scales: cell population growth in vitro predicts reduced fish growth. Sci Adv 1(7):e150030. https://doi.org/10.1126/sciadv.1500302

Streit B (1998) Bioaccumulation of contaminants in fish. EXS 86:353–387. https://doi.org/10.1007/978-3-0348-8853-0_12

Sumi N, Chitra KC (2019) Cytogenotoxic effects of fullerene C_{60} in the freshwater teleostean fish, *Anabas testudineus* (Bloch, 1792). Mutat Res Genet Toxicol Environ Mutagen 847:503104. https://doi.org/10.1016/j.mrgentox.2019.503104

Takiguchi N, Tajima T, Asayama K et al (2002) Behavioral responses of the ciliated protozoan *Paramecium caudatum* to 2,4-dichlorophenoxyacetic acid and its analogues. J Biosci Bioeng 93(4):416–420

Tamba-Berehoiu RM (2014) Small treatise on ecotoxicology. Ars Docendi, Bucharest

Tanneberger K, Knöbel M, Busser FJM et al (2013) Predicting fish acute toxicity using a fish gill cell line—based toxicity assay. Environ Sci Technol 47:1110–1119. https://doi.org/10.1021/es303505z

Teh SJ, Adams SM, Hinton DE (1997) Histopathologic biomarkers in feral freshwater fish populations exposed to different types of contaminant stress. Aquat Toxicol 37(1):51–70. https://doi.org/10.1016/S0166-445X(96)00808-9

Thanomsit C, Kiatprasert P, Prasatkaew W et al (2021) Acetylcholinesterase (AChE) monoclonal antibody generation and validation for use as a biomarker of glyphosate-based herbicide exposure in commercial freshwater fish. Comp Biochem Physiol C Toxicol Pharmacol 241:108956. https://doi.org/10.1016/j.cbpc.2020.108956

Thenmozhi C, Vignesh V, Thirumurugan R et al (2011) Impacts of malathion on mortality and biochemical changes of freshwater fish *Labeo rohita*. J Environ Health Sci Eng 8(4):393–400. https://ijehse.tums.ac.ir/index.php/jehse/article/view/315

Thomann RV (1989) Bioaccumulation model of organic chemical distribution in aquatic food chains. Environ Sci Technol 23:699–707. https://doi.org/10.1021/es00064a008

Thomann RV, Connolly JP, Parkerton TF (1992) An equilibrium model of organic chemical accumulation in aquatic food webs with sediment interaction. Environ Toxicol Chem 11:615–629. https://doi.org/10.1002/etc.5620110505

Tofan L, Niță V, Nenciu M et al (2023) Multiple assays on non-target organisms to determine the risk of acute environmental toxicity in tebuconazole-based fungicides widely used in the black sea coastal area. Toxics 11(7):597. https://doi.org/10.3390/toxics11070597

Trenfield MA, Walker SL, Tanneberger C et al (2023) Toxicity of zinc to aquatic life in tropical freshwaters of low hardness. Environ Toxicol Chem 42(3):679–683. https://doi.org/10.1002/etc.5556

Udebuani AC, Pereao O, Akharame MO et al (2023) The potential ecological risk of veterinary pharmaceuticals from swine wastewater on freshwater aquatic environment. Water Environ Res 95(1):e10833. https://doi.org/10.1002/wer.10833

Ullah S, Li Z, Zuberi A et al (2019) Biomarkers of pyrethroid toxicity in fish. Environ Chem Lett 17:945–973. https://doi.org/10.1007/s10311-018-00852-y

Ullah S, Zorriehzahra MJ (2015) Ecotoxicology: a review of pesticides induced toxicity in fish. Adv Anim Vet Sci 3:40–57. https://nexusacademicpublishers.com/uploads/files/Nexus_MH20141213011243.pdf

URL1. http://www.theplantlist.org/tpl1.1/search?q=Myriophyllum+spicatum+. Accessed 18 July 2023

URL2. https://0f10q7nd9-y-https-www-webofscience-com.z.e-nformation.ro/wos/woscc/summary/05a80529-a380-446f-9ba1-4ca44add4244-99aacc62/relevance/1. Accessed 24 July 2023

URL3. https://www.canada.ca/content/dam/eccc/migration/main/faunescience-wildlifescience/1ad45620-6a99-470c-8fdc-d57351d47ec8/rm37-202nded-lemnaenglish-20-20u.pdf. Accessed 24 July 2023

Van den Berg SJP, Maltby L, Sinclair T et al (2020) Cross-species extrapolation of chemical sensitivity. Sci Total Environ 141800. https://doi.org/10.1016/j.scitotenv.2020.141800

van der Oost R, Beyer J, Vermeulen NPE (2003) Fish bioaccumulation and biomarkers in environmental risk assessment: a review. Environ Toxicol Pharm 13:57–149. https://doi.org/10.1016/s1382-6689(02)00126-6

van Weert S, Redondo-Hasselerharm PE, Diepens NJ et al (2019) Efects of nanoplastics and microplastics on the growth of sediment-rooted macrophytes. Sci Total Environ 654:1040–1047. https://doi.org/10.1016/j.scitotenv.2018.11.183

Velisek J, Dobsikova R, Svobodova Z et al (2006) Effect of deltamethrin on the biochemical profile of common carp (*Cyprinus carpio* L.). Bull Environ Contam Toxicol 76:992–998. https://link.springer.com/article/10.1007/s00128-006-1016-9

Vinken M, Blaauboer BJ (2017) In vitro testing of basal cytotoxicity: establishment of an adverse outcome pathway from chemical insult to cell death. Toxicol Vitro 39:104–110. https://doi.org/10.1016/j.tiv.2016.12.004

Walker CH (2006) Ecotoxicity testing of chemicals with particular reference to pesticides. Pest Manag Sci 62:571–583. https://doi.org/10.1002/ps.1218

Walker CH (2008) Ecotoxicity testing: science, politics and ethics. Altern Lab Anim 36(1):103–112. https://doi.org/10.1177/026119290803600111

Weiss CM (1958) The determination of cholinesterase in the brain tissue of three species of fresh water fish and its inactivation in vivo. Ecology 39:194. https://doi.org/10.2307/1931863

Wichterman R (1986) The biology of *Paramecium*, 2nd edn. Plenum Press, New York
Wong DCL, Maltby L, Whittle D et al (2004) Spatial and temporal variability in the structure of invertebrate assemblages in control stream mesocosms. Water Res 38:128–138. https://doi.org/10.1016/j.watres.2003.09.012
Xie H, Tian X, Lin X et al (2023) Nanoplastic-induced biological effects in vivo and in vitro: an overview. Reviews Env Contamination 261:2. https://doi.org/10.1007/s44169-023-00027-z
Xu Y, Park SJ, Gye MC (2019) Effects of nonylphenols on embryonic development and metamorphosis of *Xenopus laevis*: FETAX and amphibian metamorphosis toxicity test (OECD TG231). Environ Res 174:14–23. https://doi.org/10.1016/j.envres.2019.04.010
Yancheva V, Velcheva I, Stoyanova S et al (2016). Histological biomarkers in fish as a tool in ecological risk assessment and monitoring programs: a review. Appl Ecol Environ Res 14:47–75. https://doi.org/10.15666/aeer/1401_047075
Yeltekin AÇ (2022) The responses of cholinergic system in the brain tissue of van fish (*Alburnus tarichi*) exposed to antifungal tebuconazole compound toxicity. Aquat Res 5(2):110–116. https://doi.org/10.3153/AR22010
Zhang J, Li C, Chen X, Li Y, Fei C, Chen J (2022) *Paramecium bursaria* as a potential tool for evaluation of microplastics toxicity. Biology 11(12):1852. https://doi.org/10.3390/biology11121852
Zingel P, Jeppesen E, Nõges T et al (2023) Changes in nutrient concentration and water level affect the microbial loop: a 6-month mesocosm experiment. Aquat Ecol 57:369–381. https://doi.org/10.1007/s10452-023-10015-z

Chapter 5
LC_{50} Determination and Probit Analysis

Arzu Uçar

Abstract In environmental risk assessment, the possible chronic effects of industrial chemicals, pharmaceuticals, cosmetics and agrochemicals on aquatic species can be estimated from acute toxicity data. The aquatic environment covers three quarters of the total earth's surface, making it one of the most threatened ecosystems. A large amount of toxicity test data is required to identify safer chemical compounds and then get them approved for marketing. There is not enough data for newly developed chemicals or compounds. In such cases, manufacturers, regulatory agencies and environmental scientists must demand and rely on computational-in silico modeling that tracks toxicity prediction of new and/or untested chemicals. Results from toxicity tests with varying durations and different aquatic species are used for aquatic toxicity databases. Prediction data depend on the quality of the model based on the accuracy and precision of the database. Therefore, this book chapter discusses the most frequently studied aquatic toxicity databases and comprehensive information to facilitate their efficient use by experimental scientists as well as computational scientists.

Keywords Environmental risk assessment · Toxicity tests · Probit analysis · LC_{50}

List of Abbreviations

EU	European Union
EC_{50}	Effective Concentration
IC_{50}	Inhibition Concentration
LC_{50}	Lethal Concentration
LOEC	Lowest Observed Effect Concentration
MOA	Modes of Action

A. Uçar (✉)
Department of Aquaculture, Faculty of Fisheries, Atatürk University, Erzurum, Türkiye
e-mail: arzuucar@atauni.edu.tr

M. Atamanalp et al., *Aquatic Toxicology in Freshwater*, Springer Water,
https://doi.org/10.1007/978-3-031-56669-1_5

NAMs	New Approach Methods
NOEC	No Observed Effect Concentration
OECD	Organization for Economic Cooperation and Development
QSAR	Quantitative Structure–Activity Relationship
TG203	Test Guideline

5.1 Introduction

Aquatic environments cover very large areas on earth. In addition to their vastness, their morphological structure, physical properties and biological content also show great diversity. In these ecosystems, it is necessary to determine the effects of newly developed chemicals as well as pollutants from the past that need to be identified. Increasing amounts of anthropogenic compounds entering the environment have led to the need for reliable and accurate acute toxicity tests (Lammer et al. 2009). The above-mentioned diversities cause diversity in research, making the researchers' work more difficult while expanding the field of study. Most of the legally accepted models for environmental risk assessment of chemicals, pharmaceuticals, biocides, additives, wastes are based on tests with vertebrate models such as rodents/fish (Scholz et al. 2014).

Fish have been used in toxicity tests since the 1800s (Penny and Adams 1863; Hunn 1989). The 20 years after World War II saw a marked increase in the development of "standardized" methods for short-term (96-h) testing focusing on lethality and the use of fish as toxicological models (Ankley and Villeneuve 2006).

Two main principles underlie all descriptive animal toxicity testing. The first is that the effects produced by the compound in laboratory animals are valid for humans when properly characterized. These priorities apply to all experimental biology and medicine. Based on dose per unit of body surface, toxic effects in humans are generally in the same range as those in experimental animals. On a body weight basis, humans are generally probably about ten times more vulnerable than experimental animals. This is much more effective in aquatic organisms whose habitat is water and toxic effects are more pronounced. The second main principle is that exposure of experimental animals to high doses of toxic substances is a necessary and valid method for the discovery of potential hazards in humans. Obtaining statistically valid results from such small [laboratory test] animal groups requires the use of relatively high doses so that the effect occurs frequently enough to be detectable. To detect such a low incidence in experimental animals, large numbers of animals are required. Therefore, there is no choice but to give high doses and then use toxicological principles when estimating risk and outcomes at low doses (Klaassen and Amdur 2013).

The cornerstone of safety assessments of pharmaceuticals, food additives, pesticides, cosmetics and chemicals is risk assessment. The process of assessing the hazardous properties of substances and the conditions of exposure is carried out to determine the likelihood of adverse effects to people or the environment and to characterize the nature of the effects that may be experienced (Derelanko 1995). The traditional risk assessment paradigm is evaluated in four main steps: hazard identification, exposure assessment, dose–response assessment and risk characterization. The hazard identification and dose–response steps in the risk assessment paradigm focus predominantly on the description of major toxic effects such as acute lethality, organ-related toxicity, birth defects and cancer, as well as the doses at which these effects occur. For this purpose, toxicity tests using rodents, rabbits, dogs and other animals are relied upon for higher organisms (Seidle and Stephens 2009).

The current hazard-risk paradigm is ambiguous in distinguishing between hazard and risk. Quantification of harm relies on toxicity test data that identify doses/exposures at which some form of harm is likely to occur under a defined set of conditions. Hazard-based approaches take into account reported test results. Risk-based approaches attempt to correct dose/exposure criteria for case-specific differences influenced by various toxicity modifying factors. They thus provide estimates for situations where harm is both very likely and very unlikely to be observed (McCarty 2012).

Because xenobiotics can be easily transported over long distances, accumulate in the natural environment, where they can undergo numerous transformations, reactions and interactions, it is often extremely difficult to assess the ecotoxic effects of these substances (Dordio and Carvalho 2013). The effects of a substance on living organisms are typically assessed by toxicological methods and an integration of both positive and negative effects of a substance on the growth/development of the organism. Biological methods used in toxicology to assess environmental pollution are divided into biosensors, biotests and bioanalytical, consisting of biomonitoring based on the analysis of passive samples of living matter and the observation of bioindicators (Traczewska 2008). The large number of living groups, as well as the rapidly increasing number of chemicals that need to be tested, is forcing aquatic toxicologists to constantly search for new solutions. Direct toxicity, which occurs when the concentration of a toxicant exceeds the average harmful concentration value, as determined by a series of toxicological tests, results from changes in environmental parameters under the influence of direct toxicity, thus causing detrimental effects on biocenosis (Traczewska 2011).

The damage caused by any chemical substance in the organism is called toxicity. The agent causing toxicity is called a toxicant. Toxicity tests are experiments designed to estimate the toxicant concentrations and exposure time required to produce an effect (Cope et al. 2004). Toxicity is species-specific. Individual differences should also be taken into account. In order to make a healthy species-wide determination, it is necessary to keep the number of subjects sufficient. Because individuals respond to the same dose of toxic substance at different levels.

In determining the effects of a toxicant, the researcher utilizes thousands of data. Some of these are data previously reported by other researchers, while others are their own findings. Toxicity bioassays are used to detect and calculate the potential toxicological effects of chemicals on organisms. These tests provide a database that can be used to assess the risk associated with a situation experienced by organisms. Several methods have been developed to assess the hazard and potential toxicity of chemicals to organisms, such as acute toxicity testing, sub-acute toxicity testing or chronic toxicity testing (Singh and Zahra 2017). Acute toxicity is usually determined by short-term exposure of fish to a range of chemical concentrations. In nature, most stressors can be considered acute stressors. It should also be taken into account that the acute effects of each toxicant will differ from its chronic effects.

Acute toxicity tests are the result of a short-term, high-intensity exposure (Tort 2011). The concentration that is lethal to 50% of the test fish is calculated and expressed as the LC50 value. Chronic toxicity is related to longer-term exposure. Compared to acute tests, the duration is longer and the test concentrations are lower. These tests can be carried out at one stage in the life cycle of the organism, or they can cover all life stages. The biology of the test subject is decisive at this stage.

For species with longer life cycles, researchers may prefer effects on the egg stage, hatching, growth and survival. The main determinant here is the life span of the test species. There are even species that are preferred as test species and whose entire life cycle is completed in a few days. In this case, responses resulting in survival/death of the individual or morphological responsen etc. are included. Chronic conditions where the intensity of the stressor is low but persistent are less common in nature. However, they are otherwise normally found in fish exposed to anthropogenic activities such as pollution or aquaculture. In both cases, whether acute or chronic, the development of the response is quite different. In acute attacks, the immune system may not be severely affected as rapid muscular, cardiovascular responses occur. Some basic immune mechanisms, such as phagocytosis, remain active and other immune mechanisms may not be significantly affected for short periods (Tort 2011). While acute toxicity testing is a basic requirement in most parts of EU chemical legislation, chronic aquatic toxicity testing may be needed when the result of acute testing indicates a risk or where long-term exposure is expected.

Toxicity tests, which vary considerably as they have very different characteristics, are conducted to determine the extent to which chemicals can harm living organisms in a controlled environment. It has four main objectives:

1. Obtain toxicity and exposure data for different chemicals;
2. Help to predict and manage the risks posed by different chemicals;
3. Help chemical regulations and environmental standards to be set
4. Classify chemicals according to how toxic they are to different species.

In ecotoxicology, acute toxicity tests are commonly performed to assess the harmful effects of a toxicant on an aquatic organism (Alabaster and Lloyd 1982). Such tests allow ecotoxicologists to estimate the dose–response relationship between concentrations of a given toxicant and the magnitude of their respective harmful effects on the aquatic organism selected in the tests. The median lethal concentration, or LC_{50}, is the concentration of a toxicant given at one time that causes the death of 50% of a population of test organisms. The LC_{50} is a standardized approach for assessing the short-term (acute) harmful toxicity of a toxicant (Roy et al. 2022).

Obtaining this value for each chemical is of absolute importance for the safe use of that chemical.

When determining an LC_{50} for aquatic organisms, laboratory conditions must meet standard conditions. A detail that will be ignored here will mislead subsequent researchers and will cause negativity for nature and living things during the use phase. Therefore, researchers should consider this stage as an ethical concern and think about the long-term consequences of the data they will obtain. As toxicity measurements vary by test organism and test protocol/conditions, there will also be any inherent/intrinsic determination. Similarly, for the same test organisms under various field conditions, significant differences in the nature and magnitude of various toxicity modifying factors cause variability in adverse effects (McCarty et al. 2020).

Legal requirements exist to protect wildlife and human health from the effects of toxins released into the environment. These regulations are based on the results of toxicologists' work. Researchers therefore have a great responsibility. To obtain these data, researchers take into account quantitative environmental quality criteria. Toxicity test data are often analyzed in the development of such criteria. Regression-based estimators are often used to calculate the concentration associated with a given reduction/inhibition in response relative to control conditions. Unconventional response models, often characterized as 'hormesis', can lead to alternative definitions of the underlying responses for inhibition concentration calculations.

Assessment of the harmfulness of chemical substances occurring in surface water, sewage, leachate or drinking water, among others, is carried out using acute toxicity tests. Typically, changes in the number and condition of bioindicator organisms are observed within 96 h after a single dose of the test substance. In addition, chronic toxicity tests can offer the possibility to assess the adverse effects of a toxicant not only on individuals, but also on the entire population. During long-term toxicity tests, organisms are exposed to at least five concentrations of the test substance over the entire life cycle. The main toxicity parameter is considered to be the dose, which is the amount of the substance administered, ingested or absorbed by the body that causes specific effects (Kozak and Włodarczyk-Makuła 2016; Piwowarska and Kiedrzyńska 2022).

The fish acute toxicity test is a frequently used test for chemical hazard characterization (Burden et al. 2020) and has gained a strong global presence in the risk assessment of chemicals. Fish are exposed to the test chemical for 96 h, with mortality and abnormalities recorded after 24, 48, 72 and 96 h. A concentration value (LC_{50}) is obtained that kills 50% of the fish. It was first adopted in 1981 in the Organization for Economic Cooperation and Development (OECD) test guideline (TG203). The idea that altered water chemistry could be hazardous to fish physiology was reported many years ago (Ringer 1884) and the first proper reports of toxicity experiments were established in the 1920s and 1930s. However, it was not until the 1950s that a standardized laboratory test method for the evaluation of industrial effluents for fish toxicity and new chemicals was developed. Acute testing methodology was further developed in the 1960s and early studies addressed basic issues related to oxygen requirements and aeration rates in fish, such as water aeration, apparatus design for dosing, test duration and how to report median survival time (Katsiadaki et al. 2021).

It is particularly important to determine toxicological values and effect concentrations such as LCx, ECx, NOEC or LOEC (Ducrot et al. 2014). These not only provide a scale value for further research, but also form the basis for regulations worldwide. When comparing results from different laboratories, there are sometimes differences between analyses, even with similar methodologies (Giusti et al. 2013). These deviations, which can be attributed to factors such as genetic variation, complicate the work of toxicologists. Therefore, it is essential that the data provided are thoroughly analyzed (Kaur and Dua 2015). With the preference of standardized methods, accredited laboratories and the experience of toxicologists, the results are more standardized.

Over the last 10 years, the public, private, and non-profit sectors have agreed that hazard identification and risk assessment (in vitro and in silico methods—collectively referred to as new approach methods (NAM's)) should utilize the biological sciences. The objectives for the application of NAMs are to accurately identify potential chemical hazards and collect dose–response data for human safety assessment. The development of NAMs and traditional testing approaches have been evaluated at four levels. Level 1 is based on computational screening only. Level 2 consists of high-throughput in vitro screening with human cells aimed at providing broad coverage of possible responses. Level 3 focuses on fit-for-purpose assays selected based on hypothetical modes of action (MOA) and designed to provide more quantitative estimates of relevant dose responses. Level 4 has more complex multidimensional or multicellular analyses and may include targeted in vivo studies to further define the MOA. Each level also includes exposure assessment tools appropriate to the decision. The aim of these studies is (1) to learn about NAMs in relation to risk assessment and (2) to define a functional roadmap to ensure that NAMs are adequate for chemical safety decision-making (Andersen et al. 2019).

However, understanding of the toxicity footprint of chemicals on ecosystems is hampered by the diversity of chemicals used by society, the wide range in their sensitivity across species, and the high costs of experimental toxicity data (Martin 2019). Different species may exhibit different sensitivities to the same chemical. For example, pyrethroid insecticides are highly toxic to insects but tolerated by most mammals (Wolansky and Harrill 2008). One approach to overcome the paucity of experimental data is the use of Quantitative Structure–Activity Relationship (QSAR) models.

QSARs have several potential uses, some of which include (1) predicting the effects of newly synthesized chemicals, (2) prioritization of targeted chemicals for more detailed toxicity testing, (3) assistance in deriving water quality guidelines, and (4) rapid toxicity prediction of compounds for specific situations (when toxicity test data are not available) (Blaisea and Férard 2005).

QSAR models reveal that fish or cell toxicity is closely related to chemical hydrophobicity, ionization, energy of the lowest empty molecular orbital, hydrogen bonding potential and stability. These descriptors reflect chemical bio-uptake, reactivity and interaction with target receptors. On the other hand, the binomial model reveals that mitochondrial toxicity is closely related to chemical hydrophobicity and polarizability/dipolarity. This suggests that biological uptake and Van der Waals interaction play important roles in mitochondrial toxicity. Theoretical equations were used to explain toxicity correlation, excess toxicity and QSAR in fish, cell and mitochondrial toxicity. The above results suggest that although some compounds share the same modes of action between fish-cell toxicity and mitochondrial toxicity, cytotoxicity can represent fish toxicity, can be used in safety assessment of organic pollutants in aquatic environment and mitochondrial toxicity. QSAR models can predict the bioactivity or toxicity of a chemical using the structure of the chemical in the absence of experimental data (Cherkasov et al. 2014) and generally use linear regression or logistic regression models (Chen et al. 2012). Quantitative structure–activity relationships (QSARs), which aim to predict the toxic properties of organic compounds, are based on the fact that molecules can enter living organisms and exert adverse effects depending on their elemental composition and structure. Briefly, QSARs are regression equations that relate physicochemical properties to toxicological endpoints (e.g. LC50, EC50, IC50, NOEC) within a class of compounds. Several methods have been used to investigate MOAs of compounds between two toxicity endpoints (e.g. mitochondrial, cell–fish toxicity) (Li et al. 2018). One method is the correlation between two toxicity endpoints (Bale et al. 2014). Theoretically, if the MOAs of many chemicals differ between two toxicity endpoints, the toxicity correlation is unlikely to be significant. Some in vitro-in vivo toxicity correlations have been determined to predict in vivo outcomes based on in vitro data (Bruckner et al. 2018). A positive correlation between cytotoxicity and fish toxicity was observed for a wide range of compounds with different modes of action and physicochemical properties (Huang et al. 2019). Mayer et al. (1994) estimated the chronic toxicity of chemicals to multiple fish species using a linear regression model and acute toxicity test data, and Raevsky et al. (2008) estimated the LC50 values of chemicals to Guppy (*Poecilia reticulata*), Fathead minnow (*Pimephales promelas*) and Rainbow trout (*Oncorhynchus mykiss*) using a chemical similarity approach. Quantitative structure–activity relationship (QSAR) can be used to investigate MOAs in inert, less inert, reactive and specifically active compounds. In general, the toxicity values of inert and less inert compounds to aquatic organisms are linearly related to the octanol/water partition coefficient (log KOW). For example, many QSARs are determined by the octanol–water coefficient (Kow), a well-known predictor of a compound's tendency to bioaccumulate. The lipophilicity of a compound is most commonly expressed as the octanol–water partition coefficient (Kow). The higher the log Kow value, the higher the lipophilicity value (Amézqueta et al. 2020). The octanol–water partition coefficient (KOW, typically expressed in logarithmic form as log KOW), defined as the ratio of the concentration of a chemical in the octanol phase to its concentration in the aqueous phase at equilibrium at 25 °C, is a well-known property in environmental science (Zhu et al. 2022). log KOW represents a measure of the tendency of a chemical to migrate from the aqueous phase to lipids. There are some theorized findings here. These are (1) positive values of log KOW indicate the hydrophobic property of compounds; (2) larger values indicate greater hydrophobicity of the compounds; and (3) Log KOW value greater than 3 indicates that this chemical is considered very hydrophobic (Cumming and Rucker 2017; Zhu et al. 2022).

Please devide the text on several sections with numbered titles.

5.2 Conclusion

Hazard assessment of chemicals to aquatic organisms requires the use of many toxicity tests. Acute toxicity tests measuring survival, growth and reproductive effects, embryo-larval toxicity tests and chronic toxicity tests now provide the most relative utility for assessing potential chemical hazards to aquatic life. Physiological, biochemical and histological measurements have a low relative utility as diagnostic tests in aquatic toxicology. This is because it is not yet possible to link changes in these non-lethal responses to adverse environmental effects. The problem of interpreting the toxicological significance of chemical-induced changes in biochemical and physiological mechanisms is twofold: (1) the understanding of physiological and biochemical regulatory mechanisms in fish is limited; and (2) parallel changes in these mechanisms are difficult to associate with to toxicant exposure and impaired fish survival. To overcome this problem, more physiological and biochemical investigations should be carried out in conjunction with toxicity studies that measure the full range of animal responses. At this stage, the discipline of medical toxicology offers an important opportunity for aquatic toxicologists. In the early years of aquatic toxicology, test methods and terms from this field were utilized significantly. If relevant diagnostic tests are to be developed for use in aquatic toxicology, the biochemical and physiological responses caused by the toxicant should be clearly linked to responses related to reproduction, growth and development, survival or fish health. The use of diagnostic tests in hazard assessment procedures can reduce the time required for safety assessment of chemicals, more adequately define ineffective exposure concentrations and provide a better understanding of the mode of action of chemicals.

Issues to improve the "state of the art" status of diagnostic or clinical testing in aquatic toxicology continue to be debated (Mehrle and Mayer 1980). Aquatic toxicology, which likes and absolutely needs multidisciplinary work, benefits from developments in every discipline it is involved with.

Acknowledgements Many thanks to my colleagues from Fisheries Faculty who supported me in the research for their help Atatürk University for their support in my academic studies.

References

Alabaster JS, Lloyd R (1982) Water quality criteria for freshwater fish. 2nd edn. Butterworth Scientific, pp 361. https://doi.org/10.1016/C2013-0-04159-X

Allison TC (2016) Application of an artificial neural network to the prediction of OH radical reaction rate constants for evaluating global warming potential. J Phys Chem B 120(8):1854–1863. https://doi.org/10.1021/acs.jpcb.5b09558
Amézqueta S, Subirats X, Fuguet E, Rosés M, Ràfols C (2020) Octanol-water partition constant. Liquid-Phase Extraction 183–208.https://doi.org/10.1016/B978-0-12-816911-7.00006-2
Anderse ME, McMullen PD, Phillips MB, Yoon M, Pendse SN, Clewell HJ, Clewell RA (2019) Developing context appropriate toxicity testing approaches using new alternative methods (NAMs). ALTEX-Alternatives to Animal Experiment 36(4):523–534. https://doi.org/10.14573/altex.1906261
Ankley GT, Villeneuve DL (2006) The fathead minnow in aquatic toxicology: past, present and future. Aquat Toxicol 78(1):91–102. https://doi.org/10.1016/j.aquatox.2006.01.018
Bailer AJ, Oris JT, See K, Hughes MR, Schaefer R (2003) Defining and evaluating impact in environmental toxicology. Environmetrics: The Official J Int Environmetrics Soc 14(2):235–243. https://doi.org/10.1002/env.580
Bale AS, Kenyon E, Flynn TJ, Lipscomb JC, Nedrick DL, Hartung T, Patton GW (2014) Correlating in vitro data to in vivo findings for risk assessment. https://doi.org/10.14573/altex.1310011
Blaise C, Férard JF (Eds) (2005) Small-scale freshwater toxicity investigations. https://doi.org/10.1007/1-4020-3553-5
Bruckner DM, Connerney JJ, Dordick JS (2018) Advancing in vitro–in vivo toxicity correlations v ia high-throughput three-dimensional primary hepatocyte culture. AIChE J 64(12):4331–4340. https://doi.org/10.1002/aic.16442
Burden N, Benstead R, Benyon K, Clook M, Green C, Handley J, Hutchinson TH (2020) Key opportunities to replace, reduce, and refine regulatory fish acute toxicity tests. Environ Toxicol Chem 39(10):2076–2089. https://doi.org/10.1002/etc.4824
Chen B, Sheridan RP, Hornak V, Voigt JH (2012) Comparison of random forest and pipeline pilot naive bayes in prospective QSAR predictions. J Chem Inf Model 52(3):792–803. https://doi.org/10.1021/ci200615h
Cherkasov A, Muratov EN, Fourches D, Varnek A, Baskin II, Cronin M, Tropsha A (2014) QSAR modeling: where have you been? Where are you going to? J Med Chem 57(12):4977–5010. https://doi.org/10.1021/jm4004285
Cope WG, Leidy RB, Hodgson E (2004) Classes of toxicants. Use classes. In: Hodgson E (ed) A textbook of modern toxicology. 3rd ed. Wiley and Sons, Inc., Hoboken, New Jersy, pp 49–73.
Cumming H, Rücker C (2017) Octanol–water partition coefficient measurement by a simple 1H NMR method. ACS Omega 2(9):6244–6249. https://doi.org/10.1021/acsomega.7b01102
Derelanko MJ (1995) Risk assessment
Devillers J (2001) A general QSAR model for predicting the acute toxicity of pesticides to Lepomis macrochirus. SAR QSAR Environ Res 11(5–6):397–417. https://doi.org/10.1080/10629360108035361
Dordio AV, Carvalho AJP (2013) Organic xenobiotics removal in constructed wetlands, with emphasis on the importance of the support matrix. J Hazard Mater 252:272–292
Ducrot V, Askem C, Azam D, Brettschneider D, Brown R, Charles S, ... Lagadic L (2014) Development and validation of an OECD reproductive toxicity test guideline with the pond snail Lymnaea stagnalis (Mollusca, Gastropoda). Regul Toxicol Pharmacol 70(3):605–614
Giusti A, Leprince P, Mazzucchelli G, Thomé JP, Lagadic L, Ducrot V, Joaquim-Justo C (2013) Proteomic analysis of the reproductive organs of the hermaphroditic gastropod Lymnaea stagnalis exposed to different endocrine disrupting chemicals. Plos one 8(11):e81086
https://doi.org/10.48550/arXiv.1605.01156
Huang T, Zhu D, Yang Y, Huang Y, Zhang SN, Qin WC, Zhao YH (2019) Theoretical consideration on the prediction of in vivo toxicity from in vitro toxicity: Effect of bio-uptake equilibrium, kinetics and mode of action. Chemosphere 221:433–440. https://doi.org/10.1016/j.chemosphere.2019.01.062

Hunn JB (1989) History of acute toxicity tests with fish, 1863–1987. Investigations in fish control (USA)

Kar S, Leszczynski J (2021) Aquatic toxicology databases. Chemometrics and Cheminformatics in Aquatic Toxicol 513–526.https://doi.org/10.1002/9781119681397

Katsiadaki I, Ellis T, Andersen L, Antczak P, Blaker E, Burden N, Jacobs MN (2021) Dying for change: A roadmap to refine the fish acute toxicity test after 40 years of applying a lethal endpoint. Ecotoxicol Environ Saf 223:112585. https://doi.org/10.1016/j.ecoenv.2021.112585

Kaur R, Dua A (2015) 96 h LC50, behavioural alterations and histopathological effects due to wastewater toxicity in a freshwater fish *Channa punctatus*. Environ Sci Pollut Res 22:5100–5110. https://doi.org/10.1007/s11356-014-3710-1

Klaassen CD, Amdur MO (eds) (2013) Casarett and Doull's toxicology: the basic science of poisons (Vol. 1236. McGraw-Hill, New York, pp 189–190

Kozak J, Włodarczyk-Makuła M (2016) Ogólna charakterystyka metod biologicznej kontroli jakości środowiska. LAB Laboratoria, Aparatura, Badania 21(6):22–27

Lammer E, Carr GJ, Wendler K, Rawlings JM, Belanger SE, Braunbeck T (2009) Is the fish embryo toxicity test (FET) with the zebrafish (*Danio rerio*) a potential alternative for the fish acute toxicity test? Comp Biochem Physiol C: Toxicol Pharmacol 149(2):196–209. https://doi.org/10.1016/j.cbpc.2008.11.006

Li JJ, Zhang XJ, Yang Y, Huang T, Li C, Su L, Cronin MT (2018) Development of thresholds of excess toxicity for environmental species and their application to identification of modes of acute toxic action. Sci Total Environ 616:491–499. https://doi.org/10.1016/j.scitotenv.2017.10.308

Liu Y, Racah E, Correa J, Khosrowshahi A, Lavers D, Kunkel K, Wehner M, Collins W (2016) Application of deep convolutional neural networks for detecting extreme weather in climate datasets. In Int'l Conf. on Advances in Big Data Analytics | ABDA'16 |, 81–88.

Macek KJ, Birge W, Mayer FL, Buikema AL, Maki AW (1978) Discussion session synopsis of these of aquatic toxicity tests for evaluation of the effects of toxic substances. In: Cairns J, Dickson KL, Maki AW (eds) Estimating the Hazard of Chemical Substances to AquaticLife (ASTMSTP657). American Society for Testing and Materials, Philadelphia, pp 27–32

Martin OV, Adams J, Beasley A, Belanger S, Breton RL, Brock T, Ågerstrand M (2019) Improving environmental risk assessments of chemicals: steps towards evidence-based ecotoxicology. Environ Int 128.https://doi.org/10.1016/j.envint.2019.04.053

Martin TM, Young DM (2)001 Prediction of the acute toxicity (96-h LC50) of organic compounds to the fathead minnow (*Pimephales promelas*) using a group contribution method. Chemical Research in Toxicology, 14(10), 1378–1385. https://doi.org/10.1021/tx0155045

Mayer FL, Krause GF, Ellersieck MR, Lee G, Buckler DR (1994) Predicting chronic lethality of chemicals to fishes from acute toxicity test data: concepts and linear regression analysis. Environ Toxicol Chem: An Int J 13(4):671–678

McCarty LS, Borgert CJ, Burgoon LD (2020) Evaluation of the inherent toxicity concept in environmental toxicology and risk assessment. Environ Toxicol Chem 39(12):2351–2360. https://doi.org/10.1002/etc.4881

McCarty LS (2012)"Model validation in aquatic toxicity testing: Implications for regulatory practice." Regulatory Toxicology and Pharmacology 63, no. 3: 353–362. https://doi.org/10.1016/j.yrtph.2012.04.009

Mehrle PM, Mayer FL (1980) Clinical tests in aquatic toxicology: state of the art. Environ Health Perspect 34:139–143. https://doi.org/10.1289/ehp.8034139

Penny C, Adams C (1863) Fourth report of the royal commission on pollution in Scotland. London, 2(377), 1863.

Pillai SK, Kobayashi K, Mathai T, Michael M (2021) Use of LC_{50} in aquatic regulatory toxicology-Disharmony in global harmonization of hazard classification of chemicals. Ecotoxicology and Environmental Contamination 16(1):91–96. https://doi.org/10.5132/eec.2021.01.12

Piwowarska D, Kiedrzyńska E (2022) Xenobiotics as a contemporary threat to surface waters. Ecohydrol Hydrobiol 22(2):337–354. https://doi.org/10.1016/j.ecohyd.2021.09.003

Raevsky OA, Grigor'ev VY, Weber EE, Dearden JC (2008) Classification and quantification of the toxicity of chemicals to guppy, fathead minnow and rainbow trout: Part 1. Nonpolar narcosis mode of action. QSAR Comb Sci 27(11–12):1274–1281

Ringer S (1884) Concerning the influence of saline media on fish, etc. J Physiol 5(2):98. https://doi.org/10.1113/jphysiol.1884.sp000154

Roy S, Karmakar D, Pal S (2022) Acute Toxicity Bioassay and Determination of LC50 of Cadmium Chloride in *Trichogaster* (Colisa) *fasciata*. Appl Biochem Biotechnol 194(9):3890–3900. https://doi.org/10.1007/s12010-022-03953-1

Scholz S, Ortmann J, Klüver N, Léonard M (2014) Extensive review of fish embryo acute toxicities for the prediction of GHS acute systemic toxicity categories. Regul Toxicol Pharmacol 69(3):572–579. https://doi.org/10.1016/j.yrtph.2014.06.004

Seidle T, Stephens ML (2009) Bringing toxicology into the 21st century: A global call to action. Toxicol in Vitro 23(8):1576–1579. https://doi.org/10.1016/j.tiv.2009.06.012

Singh A, Zahra K (2017) LC_{50} assessment of cypermethrin in Heteropneustes fossilis: Probit analysis. Int J Fisheries Aquatic Studies 5(5):126–130

Song R, Li D, Chang A, Tao M, Qin Y, Keller AA, Suh S (2022) Accelerating the pace of ecotoxicological assessment using artificial intelligence. Ambio 51(3):598–610. https://doi.org/10.1007/s13280-021-01598-8

Tort L (2011) Stress and immune modulation in fish. Dev Comp Immunol 35(12):1366–1375

Traczewska TM (2008) Metody biologiczne w kontroli jakości wody. Cytowane w pracy zbiorowej pod redakcją Barbary Kołwzan i Kazimierza Grabasa: ekotoksykologia w ochronie środowiska. PZITS, Oficyna Wydawnicza Politechniki Wrocławskiej, pp 435–442

Traczewska TM (2011) Biologiczne metody oceny skażenia środowiska. Oficyna Wydawnicza Politechniki Wrocławskiej.

Wolansky MJ, Harrill JA (2008) Neurobehavioral toxicology of pyrethroid insecticides in adult animals: a critical review. Neurotoxicol Teratol 30(2):55–78. https://doi.org/10.1016/j.ntt.2007.10.005

Zhu M, Su H, Bao Y, Li J, Su G (2022) Experimental determination of octanol-water partition coefficient (KOW) of 39 liquid crystal monomers (LCMs) by use of the shake-flask method. Chemosphere 287:132407. https://doi.org/10.1016/j.chemosphere.2021.132407

Chapter 6
Sampling Methods in Aquatic Toxicology

Muhammed Atamanalp

Abstract Water is essential to life, but clean water is even more essential. Therefore, many disciplines are working on the protection of clean water and water pollution. Aquatic toxicology has become more prominent in specializing in this field with its studies in recent years. The task of aquatic toxicology is to examine the substances that make up the composition of water or are subsequently incorporated into it and how they affect aquatic life.

Keywords Test organisms · Toxicology tests · Acclimatization · Aquatic insects · Pre-study

6.1 Introduction

The chemical industry is constantly developing new products for the benefit of humanity. The rapid development of this industry has accelerated the increase in product diversity. Each new chemical produced emerges as a new subject of study for toxicologists. Considering the sub-departments of toxicologists, newly developed chemicals are also emerging as study topics for aquatic toxicologists and they are trying to examine their effects on aquatic organisms. In addition, these efforts also have an imperative dimension for the commercialization of new products.

In addition to the high chemical diversity, the diversity of aquatic ecosystems also makes the work of aquatic toxicologists difficult. Considering the aquatic species living in these different ecosystems, the reason for the increasing number of researchers in this field will be understood. The high species diversity makes it difficult for researchers to determine and apply the tests they will apply to these

M. Atamanalp (✉)
Department of Aquaculture, Faculty of Fisheries, Atatürk University, Erzurum, Türkiye
e-mail: mataman@atauni.edu.tr

M. Atamanalp et al., *Aquatic Toxicology in Freshwater*, Springer Water,
https://doi.org/10.1007/978-3-031-56669-1_6

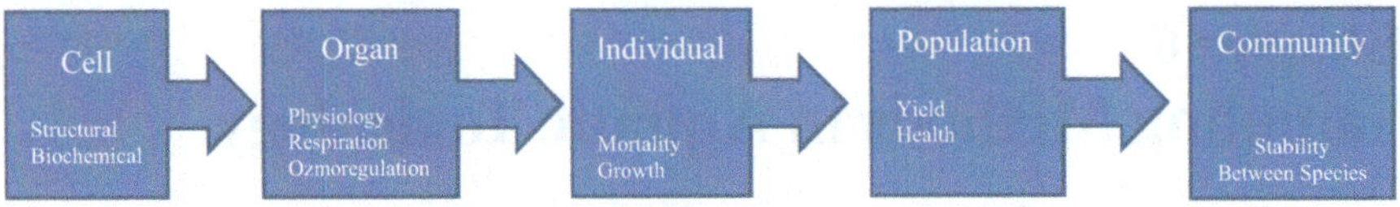

Fig. 6.1 Different levels of the organisms for determining the effects of chemicals on fish

species. Different species require different tests for toxicity tests. Although this situation makes it difficult to standardize toxicity tests, it requires the use of tests accepted by the scientific world in order for the results to have scientific properties.

At this stage of toxicology testing, experimental organisms come into play. These are commonly called bioindicator organisms and are used to biomonitor the effects of pollutants.

Measurements to determine the effects of chemicals on fish can be made at different levels of the organism. The lowest level at which measurements are made is the cell and the highest is the community, which is a collection of different populations (Fig. 6.1). Between these two extremes are organs, individuals and populations (Rand 1995; Nikinmaa 2014).

For a better understanding of toxic mechanisms in aquatic organisms, studies need to go down to the lowest level in this flow, the "cell". Each organization has its own different types of reactions. For example, there are many reactions that occur in each cell. Cells come together to form tissues and organs. These tissues and organs come together to form the fish. In a fish population, there are individuals at different life stages with different sensitivities to pollution. Therefore, the interactions of all individuals in the population are different from each other. When it comes to communities, it should be taken into account that they are formed by populations with different sensitivities in the same way. When all the interactions within each level and between levels of organization are considered together, a very complex situation emerges (Atamanalp 2019).

Investigation of metabolic and enzymatic activities in aquatic organisms is important because it allows understanding the ecological effects of pollutants. Enzymes, which are physiological and biochemical markers, are used to determine environmental pollution in aquatic organisms (Jimene 1990; Atamanalp et al 2021).

Some organisms represent an effective tool in monitoring programs for the assessment and monitoring of environmental quality in both freshwater and marine environments. Of these organism groups, those that are easier to sample and give more effective results are preferred (Atamanalp et al 2023). The selection of species to be used in aquatic toxicology experiments constitutes the most vital point of these researches. These living groups range from plankton to fish. In toxicity tests, the test material should not be limited only to living material. Other materials such as water and feed are also included in these trials.

The issues to be considered in the selection of live materials to be used in the experiments are as follows (Atamanalp 2020):

1. To be able to live in the water where the trial will be established

2. Sufficient number of individuals to be available for the tests
3. Be able to be kept in healthy conditions in the laboratory/experimental environment for at least 1 month (activity, nutrition, no lesions, etc.)
4. The biology and water parameter requirements of aquatic experimental animals should be known and provided.
5. The animals to be used in the experiment should be of appropriate size and weight.

Whichever test method is preferred for an aquatic toxicology trial or whatever type of study is being conducted, the recommendation for researchers is to conduct a so-called pilot or pre-study before the main study. Conducting a pilot / pre-study is:

1. Provides important information about the course/condition of the test organism in the environment, even if its biology is known.
2. It reveals possible errors in the calculated concentrations of the test chemical material to be tested in advance.
3. It is important to decide on the most appropriate method for the application of the test chemical.
4. It is checked that the experimental setup is working properly. The high fluidity of the water, which is the experimental medium, will not forgive even the smallest error in the setup and will cause problems during the experiment.
5. Researchers gain important experience.
6. All kinds of problems to be experienced in the main study are prevented.

6.2 Aquatic Organisms in Toxicology Experiments

Aquatic organisms which are suitable for aquatic toxicology experiments are divided into 2 groups:

1. Aquatic vertebrates
2. Aquatic invertebrates

6.2.1 Aquatic Vertebrates

Although aquatic vertebrates are a very large group, the Pisces group is most commonly used in aquatic toxicology. Fish to be collected for experiments must be absolutely healthy. They should not be stressed as much as possible during collection and transportation. Attention should be paid to sudden changes in temperature and water quality.

Fish constitute the most important group of experimental organisms due to their economic importance. In addition, the biology of these organisms and their responses to pollution are very well known. In this respect, they are the most preferred test organisms in the vertebrate group. In addition to these advantages, the following disadvantages should also be considered:

- They are sensitive to stress
- There are some difficulties in their capture and transportation to the experimental environment.
- Extensive laboratory facilities are needed for the establishment of experimental conditions and during the experiment.

Freshwater fish to be used in toxicology trials should be obtained from farms whenever possible. Legal limitations should be taken into consideration when sampling from natural environments. Particular attention should be paid to the absence of pancreatic necrosis, frunkulosis and rotational disease in cultured fish. It is important to take into consideration that fish from marine and estuarine fishes experience different life stages at different depths when collecting fish material. Since the larvae and adults of many marine fishes are extremely sensitive, care must be taken when collecting, transporting and sorting them.

Fish brought to the experimental environment are stocked in aquariums, tanks or ponds according to their species. Experimental fish are fed with natural or commercial feeds during the adaptation period. The type and size of the fish should be taken into consideration in feed selection. It should be ensured that the feeds to be used are not contaminated with pesticides and pollutants, especially metals. Overfeeding should not be done. During the adaptation period, the fish should be monitored for stress, disease, physical injuries and deaths. If any, death or abnormal individuals should be removed from the environment. If mortality exceeds 10% during adaptation, the experimental material should be renewed.

Procedures such as weighing and measuring should be carried out as soon as possible and in a way that will not harm the fish. If needed, the fish should be sedated with anesthetic substances. Considering the purpose of the experiment, the fish should be of the same size-age-sex as much as possible. The length of the largest fish should be 1.5 times the length of the smallest fish. The acclimation period is generally accepted as 14 days.

Unexpected deaths of test organisms may have occurred for many different reasons. Dense stocking according to the tank volume will prevent growth as well as disease and parasite problems.

Pathogens and parasites that are not a problem in natural waters can be a problem in intensive stocking. Unconsumed feed residues, feces and other pollution agents are potential sources of disease, bacterial growth and toxic products. Inadequate water circulation in the tanks creates dead ones in the tank.

Filtration of the water, regular cleaning and disinfection of the tools and equipment used are important before and during the trial. Stressed and weakened organisms, especially those exposed to contaminants, will be more susceptible to parasites and diseases. Feeding of fish is not recommended for short-term tests.

Fish species used in aquatic toxicology are presented in Table 6.1.

Table 6.1 Freshwater fish species used in pollution experiments

Salmonidae:	Centrarchidae:
Coregonus artedii	*Lepomis macrochirus*
Coregonus clupeaformis	*Micropterus dolomieui*
Prosopium williamsoni	*Micropterus salmoides*
Oncorhynchus gorbuscha	*Pomoxis annularis*
Oncorhynchus keta	*Pomoxis nigromaculatus*
Oncorhynchus kisutch	Percidae:
Oncorhynchus mykiss	*Perca flovescens*
Oncorhynchus nerka	*Stizostedion canadense*
Oncorhynchus tschawytscha	Percichthyidae:
Salmo clarki	*Morone chrysops*
Salmo salar	Osmeridae:
Salmo trutta	*Osmerus mordax*
Salvelinus fontinalis	
Salvelinus namaycush	
Cyprinidae:	Ictaluridae:
Carassius auratus	*Ictalurus melas*
Cyprinus carpio	*Ictalurus natalis*
Notropis atherinoides	*Ictalurus nebulosus*
Notemigonus crysoleucas	*Ictalurus punctatus*
Pimephales notatus	
Pimephales promelas	
Cyprinodontidae:	Catostomidae:
Jordanella floridae	*Catostomus commersoni*
Poeciliidae:	Esocidae:
Gambusia affinis	*Esox lucius*
Poecilia reticulata	
Clupeidae:	
Alosa pseudoharengus	
Dorosoma petenense	

6.2.2 Aquatic Invertebrates

Daphnia are of particular importance in aquatic toxicology (Egan 2023).

Daphnia:

Class: Crustaceo
Order: Cladocero
Family: *Daphnia pulex*

They are freshwater microcrustaceans, often called water fleas. Magna occur in stagnant parts of ponds, lakes and streams. Throughout the year, the majority of the Daphnia population consists of females, while the number of male individuals increases in spring and fall.

Pollution studies are carried out in 4 life stages of Daphnia (egg, juvenile, young, adult).

They are widely used in acute and chronic testing (Reilly et al. 2023). Because

- They are easily obtained
- Adapt quickly to laboratory conditions.
- Can be stocked in small volumes.
- They have short life cycles.
- They are the most sensitive organisms to chemicals.
- They have an important place in the food chain.

Phyto and zooplankton: Since these organisms are microscopic, small volumes are needed in experiments. This is desirable in experiments. Their life cycle is short. Their nutrients are well known. They are suitable for bioaccumulation experiments because they are at the bottom of the food chain. Disadvantages are that they are microscopic and require special tools and equipment for counting and identification.

Phytoplankton can be collected from the natural environment using scoops with an eye opening of less than 100 μ. However, since the selection and separation of species and the preparation of pure cultures requires a long time, ready-made cultures are commercially available.

Zooplankton are collected with dip nets with an eye opening larger than 100 μ. If the transportation distance is long, ventilation should be provided on the way. After being brought to the laboratory, healthy individuals are selected and taken into the experiment.

Crustaceans: They are particularly sensitive during shell exchange. Cannibalism among species should be considered. Adult balanus are collected by scraping them from the rocks at low tide. Decopods are collected from natural habitats where they swim freely. Larvae grow by feeding on artemia larvae.

Annelids: This group includes freshwater oligochetes and marine polychaetes. They are suitable organisms for pollution experiments. Their larger size than plankton facilitates studies and analysis of experimental results. Polychaetes live in sediment. If the water is deep, boats and special tools are required to collect these organisms. It is also possible to collect polychaetes from the shore. The sediment samples are sieved with the help of sieves to reduce their volume and these organisms are selected. The sieving process should be done carefully as they may break apart during sieving. Another method of separating organisms from the sediment is to place the sediment in a shallow tray, cover it with water and let it rest for a while. Organisms crawling on the surface are collected with fine brushes and placed in clean water. Under the microscope, the damaged organisms are separated and studies are carried out with the remaining individuals.

6.3 Feeding in Toxicology Trials

Since feed is not the main element in a toxicology trial, feeding is done with commercial feeds used in routine aquaculture. In such trials, feeding in both the control and treatment groups;

- with the same commercial feed
- in the same quantities
- at the same time
- at the same intervals
- the same number of meals
- must be done.

Failure to pay due attention to these issues, which may seem like a minor detail for those who are new to aquatic toxicology studies, will affect the health of the study results.

Studies with feeding as the main theme in toxicology trials;

- Experimenting with feed composition.
- The amount of feed was tested.
- The number of meals tested…
- It can be set up as contamination trials using contaminated feeds.

In research where feed composition is tested:

- It can be aimed to improve the compositions by adding a new additive to existing feed formulations.
- The control group will receive feed without additives, and the experimental group(s) will receive feed with additives to be tested…

In these trials.

- Coloration in fish
- Live weight gain
- Feed evaluation rate
- Parameters such as the fish meat component.

Feed quantity trials.

In these trials different amounts of feed were given to the control and treatment groups.

The quantity is different, but;

- The content is the same…
- Produced in the same batch
- Stored under the same conditions
- There must be the same freshness of feed…

In studies in which the number of meals is tested, the same feed is given to the control and experimental groups at different intervals in different numbers of meals…

6.4 Pollution Trials Using Contaminated Feed

This type of trial is part of toxicology trials. In this type of trial, the active substance to be tested is added to commercial feeds.

This addition can make with different ways:

- Spraying the contaminant (the composition to be tested) on the feed
- Contamination of feed by contamination.
- Inclusion in the feed dough.

There are some points to be considered in the fish to be tested:

1. Buy from certified producers if possible…
2. No previous infection (pancreatic necrosis, furunculosis, etc.)
3. Not exposed to toxic substances or any pollutants.
4. Development within the normal range of its species…
5. Phenotypes that do not differ phenotypically from species traits.
6. They should be the same age.
7. The size of the fish tested should be as similar as possible, which is more important in short-term tests.
8. The largest fish should be no more than 1.5 times the size of the smallest.

6.5 Stocking of Fish for Toxicology Trials

a. The size of the test medium is decided according to the size of the fish. This may be tank, small pool or aquarium etc.
b. Water supply must be according to the requirements of the fish species.
c. Water must be with appropriate quality parameters.

6.6 Obtaining the Organisms

Different ways can be followed to obtain the species that will constitute the living material of toxicology tests:

a. Obtaining from nature: In cases where the fish to be included in the trial will be obtained from nature, the following points should be considered.

- Fish should be caught with fishing gear that will not cause injury or damage…
- The least stressful fishing gear for the fish…
- The captured fish are transported to the experimental environment in a way that does not cause stress.
- The difference between the water temperature of the individuals brought to the experimental environment and the water temperature of the environment where they will be discharged.

- Hourly temperature tolerances varying according to fish species…
- Hourly temperature values tolerated by fish at different life stages
- Temperature of water in the transported environment–temperature of water in the trial environment / hourly variation tolerance
- The effect of temperature on chemical and biological phenomena is significant
- For a 10 °C increase in water temperature, the increase in chemical and biological reactions is doubled.
- The need for oxygen in warm waters is higher than in cold waters…
- Fertilizers dissolve faster and herbicides are more active and in warm water.
- Decaying materials use up more oxygen.

b. Commercially available: It is the shortest way to obtain test individuals. Since the toxic or infection history of the organisms is not known, the company/institution from which they will be obtained must be certified.
c. Individuals raised by the study group itself: It is a possible pathway for researchers with large-scale institutional equipment. It is the most reliable as researchers are aware of the entire life cycle of the species.

6.7 Sampling of Water Insects

Hand tools can be used as the water depth is not too deep. D-shaped dip net is widely used. In small lakes and ponds, specimens are captured by pressing the scoop on the vegetation in the water or on the substrate on the ground. In streams; insects are collected with the help of manual type collectors. Insects caught on the net are collected from the water by holding it in the direction of the water flow. These two setups are used to obtain qualitative data. However, if applied with multiple rules, semi-quantitative data can be collected. Wilding or Gerking rigs are used for sampling littoral insects. They allow sampling from the ground and from certain depths in the ground. Gerking device is more effective when sampling deep-rooted vegetation. They can cause errors in density estimation as they can scatter insects on the surface of the vegetation during deployment.

There are also commercially available samplers built for this purpose or simply designed by the researcher. The first one is Surber sampler, used for quantitative sampling of benthos in rivers. Hess sampler is used where Surber samplers are used. In addition, it can also sample in still waters. In addition, Current nets designed in different structures can be fixed at different depths of rivers to measure the mobility of aquatic insects.

Freeze core sediment sampler has a protective pipe around and the pipe is driven to the desired depth. After the apparatus is placed inside, it freezes the sediment sample thanks to the nitrogen passing through it. It is removed and placed in a container with ice. It is brought to the laboratory and examined. It can sample even 1 m deep from the bottom of the river bed. A major advantage of the freeze-core sampler is that it provides opportunity for vertical stratification of sub- strate cores (Hedrick et al. 2013).

Substrate-filled trays, sheet packs, multilayer sampler These are filled with sediment and attracts colonies. Micro-organisms on the leaves create attractive environments for insects (Williams and Feltmate 2017).

Ekman trap is used for sampling of immature insects from large rivers and lakes. This equipment makes semi-quantitative sampling. It is used for sampling benthos in soft substrates where lakes or rivers are ponded. The lids open on both sides and close after taking the samples inside. It is problematic to use in cases where the ground structure is rocky etc.

Multi-core sampler is used for sampling benthos from lakes with soft sediments. Sinks into the sediment with its own weight.

Air-assisted sampler has a cylinder with an air flow. It is held over the substrate to collect organisms and detritus into the chamber. The disadvantage of the system is the need for energy for air pump. Providing energy in field conditions poses a challenge for researchers (Williams and Feltmate 2017).

Plankton nets are used for collecting species found in the water column. It is attached to the back of the boat and towed. Changes in insect density according to the time of day should be taken into account when making quantitative calculations.

Although the biology of aquatic insects is extremely complex, they usually transition to terrestrial life as adults. The life stages at which sampling is carried out therefore determines the type of equipment to be used. Two different traps are used for sampling adult insects. **Mundie pyramid trap (sampler):** Adult aquatic insects undergoing post-emergence flights can be sampled using light traps or funnel nets. Light traps are best to determine the species composition of active adults or the dispersal distances traveled laterally from the stream (Smock 2007).

Edmunds sampler: They are buoyant above water. The underwater part allows insects to enter. They are designed to attract insects.

Semi-Aquatic Light Trap

Designed for nocturnal insects (insects active at night). Light is produced with battery support. Insects are collected at night by taking advantage of the insects' orientation towards the light.

6.8 Transport of Experimental Organisms from the Collection Site to the Laboratory

Transporting experimental organisms to their environment is as important as collecting them. Containers or tanks used for transportation should be elliptical in shape, especially for mobile species, to prevent them from piling up in corners or hitting and damaging the walls of the container. Slow-moving or stationary organisms do not need much more water during transport. Aquatic plants provide suitable transport media for the transportation of such organisms. During the transport of organisms, contact with air should be prevented, that is, they should not be left out of the water and this environment should be renewed as often as possible so that oxygen can be released into the water environment in which they are located.

Water temperature should be prevented from increasing during transportation. The most important laboratory conditions that affect organisms are temperature, light and ventilation.

a. **Temperature:** In general, the laboratory temperature is higher than the temperature of the natural environment where the organisms are collected. For this reason, some methods such as ventilation system, cold water bath are needed to control the temperature. The temperature should not change more than 1–2 °C during the experiment.
b. **Light:** Light intensity and duration vary according to the light requirements of organisms. For example, phytoplankton need light for photosynthesis. Some organisms (many crustaceans and fish) require a dark/light light period. Some other organisms do not need an artificial light source if there are windows in the laboratory. If different light intensities are not required for the species to be used in the experiment, the light period can be adjusted with an electric timer according to the daily light period of the experimental area.
c. **Aeration:** Aeration is needed both during the adaptation of the organisms and during the experiment. For this purpose, small aquarium motors can be used or larger pumps can be used to distribute air with rubber pipes and hoses.

6.9 Acclimation/Acclimatization

One of the important stages in experiments is the acclimation period. Acclimation is the period of time expected for the experimental material to adapt to the environment in which the experiment will be conducted before starting toxicological procedures. Although this period is tried to be standardized, the experience and observations of the researcher are extremely important at this stage. In this period, which is important for the health of the study results, it is necessary to show the necessary patience and decide that the environment and living material are fully ready for the experiment. Acclimation, thermal, light level, stock environment, stock quantity, etc. should be considered in detail.

At this stage, confusion arises between the terms adaptation and acclimation, which seem very similar, and researchers may make mistakes in both speech and scientific writing.

The difference between adaptation and acclimatization is as follows:

Acclimation refers to a gradual reversible change to the body of an organism to better suit the changes of the environment while adaptation refers to the process by which a species becomes fitted to its environment as a result of natural selection, acting upon heritable variation over several generations. Adaptation is irreversible but acclimatization can be reversed once the previous conditions are met. Adaptation is a long-term permanent adjustment of a group of organisms to a changing environment.

Acclimatization is a short-term rapid temporary adjustment of an organism to a changing environment. Adaptation occurs over many generations. Acclimatization occurs during an organism's lifetime. Adaptation is a gradual, permanent change that is evolutionary important for a species' survival and continuity, Acclimatization is a quick, temporary change that can be reversed once the previous conditions are met.

Before starting the experiment, the following points should be taken into consideration when feeding aquatic organisms with appropriate feeds during the process of acclimatization:

- Suitable for fish mouth opening
- Has the desired protein level
- Stored under suitable conditions, without mold, fungus, etc....
- The feeds used before the trial should also be preferred during the adaptation process...
- Overfeeding should not be done

In the acclimatization process for the living materials of the experiment, these points must be considered:

- Disease should be monitored
- Stress should be monitored
- Is there physical harm?
- Was there a death? If so, in which group, on which day, estimated cause of death, details of autopsy if performed.
- If there are more than 10% deaths, change the trial population...
- Every step should be recorded, interventions made
- Care should be taken in all applications and unnecessary ones should be avoided.
- What is necessary should be completed in as short a time as possible...
- Use anesthesia for weight weighings, height measurements and other procedures...

6.10 Acclimatization (Acclimation) of Experimental Organisms to Laboratory Conditions

If pollution experiments are carried out routinely, it is helpful for researchers to have adapted organisms in the laboratory at all times. If the nature of the study requires research in non-adapted organisms, care should be taken to ensure that the experimental environment is as similar as possible to the natural conditions in which the organism was collected. These special circumstances should be taken into account when evaluating the results of experiments conducted under these conditions. When organisms collected from the natural environment are brought to the laboratory, those that appear injured, parasitized or unhealthy are separated.

The others are transferred as soon as possible to stock tanks of appropriate size. Disinfection conditions must be fulfilled before the organisms are placed in these tanks. It should be taken into consideration that disinfection applications will prolong the acclimation period. Quarantine tanks can be used at this stage. Organisms placed in these tanks are kept in quarantine for at least two days.

A sufficient number of organisms of the desired size, age and sex are taken. The duration of stay in acclimation medium depends on the type of organism and the quality of the water. In order to keep the organisms in suitable conditions, the saturation of dissolved oxygen concentration should be above 60%. Necessary precautions should be taken to prevent the accumulation of aquatic organisms in the tanks.

Sudden changes in water quality parameters, especially temperature, should be avoided to prevent stress. Changes in temperature and water quality should be gradual and aquatic organisms should be acclimatized slowly. In general, temperature changes should not exceed 1 °C in 24 h until the experimental environment temperature is reached. When the organisms reach the desired experimental temperature, they should be kept at this temperature for at least two days before starting the experiment. If possible, a longer period is preferred.

Salinity changes should not exceed 0.05% within 24 h. Sudden changes in light, such as turning the light on or off suddenly, should be avoided. If possible, a light period close to the natural environment (14 h of light, 10 h of darkness) should be applied. If seasonal or daylight studies are to be carried out, timed devices should be used.

Since the organisms are brought from the natural environment in the first days of adaptation, feeding should not be done since they will not take feed. In the following days, feeding is done once or several times a day according to the wishes of the organisms. Feeding is usually stopped 24 h before the beginning of the experiment. The tanks in which the organisms are stocked should be cleaned periodically to prevent the accumulation of unconsumed feed/metabolic waste or other organic matter.

Organisms that die or show signs of disease during adaptation should be removed from the environment. All of these should be recorded. Appropriate tools and equipment should be selected according to the type of experimental organisms. After preparing the necessary equipment and environment for the experiments, the organisms to be used should be checked whether they are healthy for the last time.

Preliminary experiments are recommended before the actual trial. These experiments are carried out to determine the concentration of the pollutant to be used in the experiments. If the pollutant to be used in the experiment is a liquid, 100, 10, 1 and 0.5% of this pollutant is taken first. If the chemical tested is metal, concentrations are prepared as 5,10,25,50 and 100 mg/l and 2–4 organisms are used for each concentration. Generally, each concentration is allowed 15–24 h. After the results of these preliminary experiments are obtained, the main experiments are started. To start the main experiments, the test vessels are first placed in a temperature-controlled environment where concentration determination experiments are performed. Adequate water supply is ensured and the tanks are numbered according to the work plan. Since there is a risk of splashing in some fish species, the tanks should be covered with suitable covers.

If ventilation is required in the middle of the experiment, the ventilation system is set up so that each test container receives equal and continuous air. Water quality parameters, especially pH and temperature, are monitored and recorded at regular intervals before and during the experiment. If the organisms to be used are of different sizes, care should be taken to place equal amounts of large and small individuals in each tank. For each experimental concentration and control group, the experimental vessel is prepared in series according to the number of replicates in accordance with the experimental setup. Increasing the number of replicates increases the labor force but contributes significantly to the health of the results. If the researcher's statistical knowledge is not sufficient while setting up the experiment, it is recommended to get support from an expert.

Separate buckets, dip nets etc. should be used for each experimental unit, even if they are in the same treatment group, and the risk of chemical or disease agent contamination from these units to each other should be eliminated. Due to the lower buffering capacity of fresh water, it is more important to monitor the pH of the water.

6.11 Anesthetic Substances

Since fish are cold-blooded creatures, they can be easily stressed by different factors. Especially in physiology and behavioral studies, stress affects the health of the research results. At this stage, anesthetics can be used to prevent stress effects. The most important issue to be considered when using anesthetics in toxicology experiments is the interaction of the test chemical and the anesthetic substance.

Anesthetic substance preference in the trials:

- The application time should be between 3 and 15 min. In other words, the anesthetic should affect the fish within 3 min and the fish should be kept for 15 min in this way.
- Recovery time after use should be 5 min or less. The fish should move within 1–2 min after being placed in the recovery tank and start swimming within 5 min.
- It should be non-toxic to fish and have a wide range of safety.
- It should not be harmful to the user in hand contact in normal use.
- It should not leave a permanent effect on fish physiology and behavior.
- It should be metabolized very quickly, leave no residue, and have no excretion period. However, it should be kept for at least 21 days in fish to be offered for human consumption.
- There should be no cumulative effect in repeated use. In repeated applications, the fish should be anesthetized in the same environment and at the same dose for the same duration.
- The anesthetic to be used must be economically viable.

Clove oil in fish is an ideal anesthetic including these properties (Uçar et al 2013). Main advantages of clove oil can be pointed out as; its regulating attribute is limited, its excreting time from the body is short, it is rather safe for human and animals, it can be provided easily and it is inexpensive. It almost meets every measure that should be considered during the selection of anesthetics (Marking et al. 1985; Keene et al. 1998; Uçar et al. 2013).

Besides this, the most important disadvantages of clove oil can be pointed out as follows; its application for a long time would affect the oxygen consumption, it causes color changes in perch, brown trout and atlantic salmon; it causes equilibrium loss for fishes when there is small concentration increases at different carriage conditions in large carrying tanks and it is hard to control the concentration (Hoskonen and Pirhonen 2004), the long time of exposure during carriage causes increase of ammonia level. It is mixed with solvents as alcohol as it is not dissolved in water and when it should be used the least dosage that produces anesthesia should be used (Kaiser et al. 2002; Uçar et al. 2013).

In addition, there are some factors that may have an effect on the anesthetic to be used:

- the general composition of the water,
- whether it is salt or fresh water,
- the hardness of the water,
- pH value and its change,
- buffer capacity,
- level of metabolic wastes in water
- most important is the water temperature.

6.12 Conclusion

On the one hand, while aquatic toxicology tests are trying to be standardized, on the other hand, the increase in the factors accepted as pollutants forces this branch of science to be dynamic. In addition, new groups of organisms are being added to the test organisms, making the researchers' work even more difficult. In this chapter, we have tried to compile the basic issues that should be considered while sampling the organisms to be used in aquatic toxicology tests and setting up experiments.

Acknowledgements Many thanks to my colleagues from Fisheries Faculty who supported me in the research for their help Atatürk University for their support in my academic studies.

References

Atamanalp M, Parlak V, Özgeriş FB, Çilingir Yeltekin A, Ucar A, Keleş MS, Alak G (2021) Treatment of oxidative stress, apoptosis, and DNA injury with N-acetylcysteine at simulative pesticide toxicity in fish. Toxicol Mech Methods 31(3):224–234

Atamanalp M, Kokturk M, Gündüz F, Parlak V, Ucar A, Alwazeer D, Alak G (2023) The use of zebra mussel (*Dreissena polymorpha*) as a sentinel species for the microplastic pollution of freshwater: the case of beyhan dam lake, Turkey. Sustainability 15(2):1422. https://doi.org/10.3390/su15021422

Atamanalp (2019) Sampling methods, Lecture notes for PhD students, Not Published
Atamanalp (2020) Aquatic toxicology. Lecture notes for Erasmus PhD students, Not Published
Nicole E, Stinson SA, Deng X, Lawler SP, Connon RE (2023) "Swimming behavior of *Daphnia magna* is altered by pesticides of concern, as components of agricultural surface water and in acute exposures. Biology 12(3):425. https://doi.org/10.3390/biology12030425
Hoskonen P, Pirhonen J (2004) The effect of clove oil sedation on oxygen consumption of six temperate-zone fish species. Aquac Res 35:1002–1005. https://doi.org/10.1111/j.1365-2109.2004.01084
Jimene BD (1990) Detoxication Enzymes as Indicators of. InAmerican fisheries society symposium, vol 8. pp 67–79
Kaiser H, Brill G, Cahill J, Collett P, Czypionka K, Green A, Orr K, Pattrick P, Woody CA, Nelson J, Ramstad K (2002) Clove oil as an anaesthetic for adult sockeye salmon: field trials. J Fish Biol 60:100–105
Keene JL, Noakes DLG, Moccia RD, Soto CG (1998) The efficacy of clove oil as an anaesthetic rainbow trout, *Oncorhynchus mykiss* (Walbaum). Aquac Res 29:89–101
Malison R, Benjamin J, Baxter C (2010) Measuring adult insect emergence from streams: the influence of trap placement and a comparison with benthic sampling. J North Amer Benthol Soc 29:647–656. https://doi.org/10.1899/09-086.1
Marking LL, Mayer PP (1985) Are better anesths needed in fisheries. Fisheries 10:2–5. https://doi.org/10.1577/1548-8446(1985)0102.0.CO;2
Nikinmaa M (2014) An introduction to aquatic toxicology. Academic Press, pp 252. ISBN: 9780124115743https://doi.org/10.1016/B978-0-12-411574-3.00001-3
Trost JJ, Christy TM, Bekins BA (2018) A direct-push freezing core barrel for sampling unconsolidated subsurface sediments and adjacent pore fluids. Vadose Zone J 17:1–10 180037. https://doi.org/10.2136/vzj2018.02.0037
Rand GM (1995) Fundamentals of aquatic toxicology, fundamentals of aquatic toxicology. 2nd edn. CRC Press, pp 1148. https://doi.org/10.1201/9781003075363
Uçar A, Atamanalp M, Çankaya M, Özdemir H (2013) Effects of anesthetic substances on some antioxidan enzyme activities of trouts. J FisheriesSciences.com 7(2):152–160. https://doi.org/10.3153/jfscom.2013016
Williams D, Feltmate BW (2017) Aquatic insects. Blackburn Press, pp 374. ISBN 1932846298

Chapter 7
Classification of Pollution and Their Entry Rotues into Aquatic Ecosystems

Veysel Parlak

Abstract Pollution is defined as a condition that occurs when any substance or form of energy exceeds the carrying capacity of the environment or can be dispersed into larger areas. The term "environmental pollution" refers to the event where foreign substances that negatively affect the living and non-living elements of the environment, create physical and biological harm, and alter their qualities, mix extensively with the air, water, and soil. The main causes of environmental pollution can be listed as rapid population growth, unplanned urbanization, industrialization, and the excessive use of natural resources. Environmental pollution is generally classified into air pollution, water pollution, soil pollution, noise pollution, and visual pollution. Environmental pollutions harm all living beings in nature directly or indirectly, as they cause damage to nature itself. Today, visual pollution and light pollution are also considered forms of environmental pollution, created by humans and causing harm to both humans and other living organisms. Additionally, environmental pollution disrupts ecosystem balances, leading to climate change and global warming on a climatic scale.

Keywords Aquatic ecosystems · Classification of pollution · Air pollution · Water pollution · Soil pollution

List of Abbreviations

DNA	Deoxyribonucleic acid
CO	Cobalt
CO_2	Carbon dioxide
H_2O	Water
H_2S	Hydrogen sulfide

V. Parlak (✉)
Department of Basic Sciences, Faculty of Fisheries, Atatürk University, Erzurum, Türkiye
e-mail: veysel.parlak@atauni.edu.tr

M. Atamanalp et al., *Aquatic Toxicology in Freshwater*, Springer Water,
https://doi.org/10.1007/978-3-031-56669-1_7

NH_3	Ammonia
PH_3	Hydrogen phosphide
SO_2	Sulfur dioxide
MEK	Methyl ethyl ketone
Cl_2	Chlorine gas
ClO_2	Chlorine dioxide
HCN	Hydrogen cyanide
HCl	Hydrochloric acid
HF	Hydrofluoric acid
NOx	Nitrogen oxides
NO_2	Nitrogen dioxide
RSH	Recreational water diseases
RNH_2	It is an organic compound containing the amino group
N	Nitrogen
P	Phosphorus

7.1 Introduction

In parallel with the rapid growth of the world's population, industrialization has caused numerous environmental problems and decreased the quality of resources in ecosystems. While, globally, the rate of economic growth has been increasing in recent years, and agricultural activities, industrialization, population growth, urbanization have begun to negatively impact life. This has a detrimental effect on all natural resources. The increasing environmental pollution day by day is causing irreversible damage to the environment. It is highly important to classify pollution sources and take necessary measures to address these issues (Lin et al. 2022).

Nature, one of the fundamental elements of the environment, possesses unique physical, chemical, and biological characteristics. Taking these characteristics into consideration, environmental pollution is divided into the following sections.

7.1.1 Physical Pollution

Environmental pollution is the occurrence in which some or all of the physical characteristics of the components of the environment—soil, water, and air—are altered and disrupted in a way that threatens or adversely affects the health of humans, animals, and plants. For example, the discharge of various factory pollutants into rivers and lakes, or the natural erosion of soil into lakes and seas, can result in a change in the color of water, ranging from light brown to red or black. This phenomenon represents the physical pollution of the waters.

7.1.2 Chemical Pollution

Chemical pollution of the natural environment is the alteration of the chemical characteristics of the components of the environment—soil, water, and air—in a way that adversely affects the vital activities and activities of living organisms. For example, the discharge of various factory solid and liquid wastes into fertile agricultural lands or into rivers and streams results in agricultural soils, rivers, and lakes being exposed to chemical pollution, including harmful heavy metals.

7.1.3 Biological Pollution

The contamination of the natural environment, consisting of soil, air, and water, by various microorganisms and, consequently, the disruption of the microbiological structure is referred to as microbial pollution, while the contamination of the same environments by microorganisms is called biological pollution. For example, irrigating agricultural fields with sewage water or discharging sewage into rivers, lakes, and seas can introduce pathogenic microorganisms present in sewage water into the soil, water, and atmosphere, leading to the microbiological pollution of these environments (Ukaogo et al. 2020).

Pollution types arise as a result of natural sources or human activities. Pollution is generally classified into categories such as Air Pollution, Water Pollution, Soil Pollution, and Noise Pollution. In addition to these common types of pollution, there are rarer but more severe types, such as Light Pollution, Thermal Pollution, and Radioactive Pollution (Tiwary and Williams 2018).

Pollutants in nature undergo processes in two different ways. Biodegradable pollutants rapidly break down through natural processes. Non-biodegradable pollutants, on the other hand, have structural characteristics that do not break down in the environment or take thousands of years to break down. The classification of pollution caused by these substances can be done as follows (Daly and Zannetti 2007):

Air Pollution

Air pollution can be described as the presence of one or more pollutants in the atmosphere that can harm humans, plants, and animals, including particles like dust, smoke, haze, and odors. The sources of air pollution can be categorized as natural pollution (resulting from volcanic eruptions, forest fires, and biological decay) and human-made activities, including thermal power plants and agricultural activities. Primary pollutant sources release harmful gases directly into the atmosphere, such as carbon monoxide (CO) and nitrogen oxides (NO). In the secondary stage, these compounds can react with each other or with the basic components of the air to form new pollutants (Brunekreef and Holgate 2002).

Water Pollution

"Water pollution can be defined as the alteration of the physical, chemical, and biological characteristics of water in a way that can have harmful effects on human and aquatic life." Water pollution can originate from infectious agents found in human and animal waste. These agents can be classified as bacteria, viruses, protozoa, and parasitic worms. Various diseases can result from this type of pollution. Additionally, water pollution can occur due to production processes resulting in the release of water-soluble organic chemicals (compounds of toxic metals such as lead, arsenic, and selenium) from sources such as sewage, paper mills, and food processing facilities. Furthermore, thermal pollution occurs as a result of water cooling and overheating processes in power plants and certain industrial facilities. In such cases, there is a decrease in the dissolved oxygen levels in the water source, making aquatic organisms more vulnerable to diseases and toxic chemicals.

Water pollution and the lack of access to clean water sources have become almost a limiting factor for humanity. Lotic (riverine) ecosystems are among the most vulnerable freshwater bodies to pollution due to their role in transporting industrial and domestic wastewater and water flow from agricultural areas. Controlling water pollution in wetlands is crucial for water quality management, as the quality of water is just as important as its availability (Chaudhry and Malik 2017).

Marine Pollution

Coastal areas encompass rich coral reefs, wetlands, and seagrass beds, serving as habitats and feeding grounds for aquatic organisms. The contamination of the sea due to the disposal of waste materials into the marine environment, which disrupts water quality and the vital activities of aquatic organisms, can be defined as marine pollution. For example, the presence of heavy metals and organic pollutants can be fatally harmful to aquatic life. Similarly, waste-induced oil films on the water surface significantly affect the oxygen uptake rate in the aquatic environment and have a negative impact on aquatic life (Landrigan et al. 2020).

Soil Pollution

Soil, like air and water, is another essential element for the survival of living beings. It serves as the primary repository of the resources that nourish plant life. The top layer of soil is a fundamental source of nutrition for humans and other organisms. Within a gram of soil, there are millions of living organisms, each of which plays a unique role in maintaining the ecosystem. The upper layer of soil, which is rich in humus, is responsible for soil fertility. Soil is one of the most important natural resources, and its misuse for non-agricultural purposes, contamination with heavy metals, and the effects of erosion result in losses and reduced productivity. Recovery of lost soil is a challenging task. Furthermore, soil pollution discharge points become water sources and directly impact aquatic ecosystems in a negative way (Saha et al. 2017).

Radioactive Pollution

Nuclear power plants contribute to a type of environmental pollution known as nuclear pollution. The increasing developments in the defense industry and the work carried out in these areas have also played a significant role in the increase of this pollution. The consequences of radioactive pollution can be severe, and it can lead to the emergence of incurable diseases (Hatra 2018).

Noise Pollution

This is a type of environmental pollution, which results from the use of auditory stimuli for various purposes and the disturbing sounds produced by vehicle engines, explosive materials like fireworks, and many similar elements, affecting all living creatures in nature in a multifaceted manner (Gupta et al. 2018).

The universe we live in continues to exist alongside elements that operate in cycles. Each element functions within its own order, and the environment is a significant part of these elements. Humans should be aware of the order of the environment they live in and act accordingly. Unfortunately, people often struggle to adhere to this simple rule. Any harm inflicted by humans on the environment in which they live constitutes environmental pollution. From the waste discharged by factories to the simple deodorants we use in personal care, from the gases emitted by land and air vehicles to small plastic bags, many things contribute to environmental pollution. The concept of environmental pollution covers a wide range of factors. In its simplest form, anything that harms nature causes environmental pollution (Chowdhary et al. 2018). Some of the most noticeable reasons for this are as follows:

- Factories not limiting their waste emissions
- Disposal of trash in nature
- Excessive use of unnecessary auditory stimuli
- Non-recycling of household and medical waste
- Excessive use of non-biodegradable plastics in the environment
- Uncontrolled deforestation and destruction of natural habitats
- Widespread use of devices emitting greenhouse gases
- Excessive use of chemical-based products
- Excessive use of radiation-emitting technological devices
- Over-illumination
- Lack of oversight of nuclear power plants.

7.2 Water Ecosystem, Water Pollution and Pollutants

Water ecosystems and/or aquatic ecosystems can be defined as environments where the habitat-providing plants (flora = plant life) and animals (fauna = animal life) interact with and adapt to the surrounding chemical and physical characteristics of the water. The water bodies on our planet, such as oceans, seas, estuaries, coral reefs, rivers, lakes, and coastal areas, are the most prominent examples of aquatic ecosystems. In these systems, numerous organisms such as macrophytes, macroinvertebrates, phytoplankton, benthic algae, and fish are members that create habitats in these ecosystems. Aquatic ecosystems are generally divided into two broad categories: "saltwater and freshwater ecosystems." Saltwater (or ocean/marine ecosystems) covers about 70% of the Earth's surface, while freshwater ecosystems make up only about 6% of the surface area (Mostofa et al. 2016).

Saltwater and freshwater ecosystems, classified into two main groups, serve many important benefits and functions, including water filtration, dilution, and storage; flood prevention; maintaining local and global climate balance; and preserving biodiversity (Vo et al. 2012).

Freshwater resources can be further classified into two main groups: surface waters (such as rivers, lakes, wetlands, and reservoirs) and groundwater. Both of these water resources are crucial for the survival of humans and other living organisms. Approximately 98% of freshwater resources are found underground. Generally, groundwater is more commonly used for drinking, domestic, and agricultural irrigation purposes in rural areas, while in large cities, surface waters (such as rivers, lakes, and reservoirs) are often tapped for water supply needs (Kamtchueng et al. 2014).

Water pollution is the result of numerous anthropogenic activities that discharge various pollutants into aquatic environments, posing a threat to the health of organisms and degrading the environmental quality of water bodies (Masindi and Muedi 2018; Bashir et al. 2020). The final destination for both solid and liquid chemicals introduced into the environment is aquatic ecosystems. Both domestic and industrial wastes, including heavy metals, detergents, microfibers, plastics, or non-plastic materials, contribute to the increase in water pollution. Aquatic environments exposed to a high influx of these substances are susceptible to the excessive toxic effects of these pollutants (Sonone et al. 2020). Some of the chemicals that reach aquatic ecosystems include

- Radioactive elements (Radon, Actinium, Thorium, Protactinium, Uranium, Radium);
- Metals (Arsenic, cadmium, mercury, lead);
- Industrial solvents and "volatile organic compounds" (CO, CO_2, H_2O, H_2S, NH_3, PH_3, SO_2, Acetone, Benzene, MEK, Toluene, Xylene, Cl_2, ClO_2, HCN, HCl, HF, NOx, NO_2, RSH, RNH_2, Butane, Diesel, Ethanol, Formaldehyde, Gasoline, Methyl Bromide, Ozone, Phenol, Trichloroethylene, Vinyl Chloride);
- Agricultural chemicals (Pesticide-Herbicide);
- Domestic products (detergents, cleaners, paints);
- Fuel combustion (N and sulfur oxides, polycyclic aromatic hydrocarbons, Carbon monoxide and carbon dioxide);
- Nanoparticles;
- Personal care products;
- Microplastics;
- Antibiotics;
- Over-the-counter drugs and pharmaceuticals (Bukola et al. 2105; Schweitzer and Noblet 2018; Borahet al. 2020).

Organisms need to exchange matter with their environment in order to sustain their lives. The movement of matter between living and non-living components of the environment is called matter cycling. The cyclical use of matters is one of the most important tasks of an ecosystem. The accumulation of organic matter in a system enhances the productivity level of that system. Soil carried into aquatic ecosystems from the surrounding land contains organic material. The transport of agricultural and domestic waste into aquatic environments leads to a decrease in oxygen levels and the formation of H_2S, negatively impacting both the environment and the organisms within it (Karri et al. 2021).

7.3 Eutrophication

Nitrogen, phosphorus, and other nutrients gradually accumulate in aquatic ecosystems. The accumulation of organic matter in a system enhances the productivity level of that system. Soil carried into aquatic ecosystems from the surrounding land contains organic material. In lakes where agricultural and domestic waste mix, the increase in nitrogen and phosphorus nutrients in the water accelerates eutrophication, leading to excessive algae growth and increased turbidity in the spring months (Schindler 2012).

Rivers serve as receiving environments for wastewater. The highest levels of pollution in a river are typically observed at the wastewater discharge points because river water flows continuously at a constant or variable rate. After the discharge of carbonaceous and nitrogenous substances into the river, biochemical reactions take place, leading to the oxidation of organic matter and a decrease in the dissolved oxygen concentration in the river (Beiras 2018).

A portion of the water that falls to the Earth's surface with rainfall begins to flow, some evaporates, some is retained by plants, and some infiltrates into the soil. In order for groundwater to form, surface waters must infiltrate underground. The water that infiltrates underground accumulates on an impermeable layer and forms groundwater. The most prominent reason for groundwater pollution is the discharge of urban (nitrates and pathogens) and industrial wastes into the environment, which is subsequently transported to groundwater depending on climate, soil structure, and time (Culver and Pipan 2014). Another significant reason for groundwater pollution is the indiscriminate use of agricultural pesticides and fertilizers (nitrate, phosphorus), as well as the direct disposal of domestic waste into the soil. Nowadays, one of the most important causes of groundwater contamination is the direct disposal of domestic waste into the soil. Compounds resistant to degradation, such as detergents, can reach groundwater and pose potential risks to drinking water. Another important pollutant is the leakage from landfill sites and some storage tanks, such as those used for petroleum (Khayat et al. 2012.).

Transport Mechanisms of Pollutants in Groundwater:

1. Diffusion: Transport of pollutants according to concentration difference
2. Advection: Transport by water velocity
3. Mechanical dispersion: Transport of pollutants in a porous medium
4. Hydrodynamic dispersion: Transport of pollutants by both diffusion and mechanical dispersion

Water Quality Criteria

Although there are many water quality parameters, the main parameters used in water quality are: Dissolved oxygen is the main determinant of water quality in streams, lakes and other aquatic environments. Oxygen demand, especially biochemical oxygen demand (BOD) and chemical oxygen demand (COD) tests are used to determine the pollution potential of the discharged wastewater. Water does not look good when it contains suspended solids (AKM). Sedimentable solids precipitate to the bottom of the aquatic environment and cause mud layer formation. In addition, dissolved solids can harm the life of the water and the people who drink this water. Nitrogen and phosphorus are useful and important parameters used in water discharge into streams and lakes. The presence of viruses and bacteria that can cause infectious diseases in water is undesirable for human health (Feng 2012).

7.4 Pollution of Aquatic Ecosystems and Its Adverse Effects

With the increasing world population and increasing shelter and food needs, requirements are increasing in every field. With developing technology and industrialization, many activities that destroy nature and produce waste (Canals, Roads, Dams, Agricultural pesticides, Domestic and Factory waste) are taking place. All of these constitute the source of water pollution.

Nutrients reach water resources from point and non-point (diffuse) pollutant sources. Point sources are domestic wastewater and industrial wastewater. Diffuse sources are fertilization in agricultural areas, leachate of solid waste landfills, bovine, poultry and ovine waste, atmospheric transport, surface runoff and erosion (Malik et al. 2020).

While the relationship between nature and humans continues to develop in a balanced way, the development problem of humankind has caused the balance in this relationship to be upset against nature. Nowadays, humans strongly affect almost every major aquatic ecosystem and, through their activities, alter the fluxes of growth-limiting elements from the soil to the receiving aquatic environment. The primary reason for this is the rapid increase in agricultural activities. Atmospheric deposition of nutrients, especially phosphorus, into surface waters has also increased. In addition, people use flowing surface waters as a safe wastewater disposal system. N and P loading to Earth's surface waters is strongly influenced by human population density and land use. Population growth and activities such as urban discharges, agriculture, wastewater treatment, and burning of fossil fuels have dramatically increased the input of nutrients into the receiving environment above the natural level (Table 7.1) (Rajaram and Das 2008; Zhou et al. 2017; Auta et al. 2017; Sharma et al. 2018; Kazour et al. 2019; Patel et al. 2020; Baloyi et al. 2021).

Most developed countries implement intensive agricultural programs to meet the food needs of the world population. Unfortunately, the most important factor in the deterioration of nature (Aquatic ecosystems) is agricultural processes. In the European Union, 38% of aquatic ecosystems are under significant agricultural pressure (Table 7.2) (Mateo-Sagasta et al. 2017).

Methods to reduce nutrient input to aquatic environments in a specific way are divided into 4 categories

(a) reduction of urban and rural sources of pollution,
(b) agricultural sources,
(c) and atmospheric loads,
(d) use of wetlands and buffering as a nutrient stopper.

Entry of nutrients into ground and surface waters from point and non-point pollutant sources should be prevented. It can be reduced by conscious fertilization in agriculture and control of pollution caused by fertilization in agricultural areas. It is very important to build domestic and industrial wastewater treatment facilities, improve existing treatment facilities, ensure strict control and optimum operation (NSTC 2003; Sağlamtimur and Sağlamtimur 2018).

7.5 How to Prevent Environmental Pollution?

Harming the environment we live in is just like what a chick does when it tries to break its egg before it completes its vital data. A broken egg will be the end of the chick, and the damaged environment will bring the end of humanity. For this reason, every individual should be prudent towards the environment they live in and avoid any behavior that will cause pollution. The duties of individuals and relevant institutions to prevent pollution can be listed as follows:

- Grouping of domestic wastes in accordance with recycling;
- Preventing waste oils from mixing with soil and water;
- Reducing the use of devices that emit radiation;
- Preferring products with minimum greenhouse gas emissions in white goods;
- Reducing the use of plastic, preferring glass bottles and cloth bags instead of plastic ones;
- Reducing the use of deodorant;
- Limitation of wastes and gases discharged from factories;
- Preferring electric vehicles instead of fuel-powered vehicles;
- Use of audible stimuli only where necessary;

Table 7.1 Pollutant sources and their discharge pathways to aquatic environments

Contamination to the natural environment	Source	Pathway
Organic pollutants	Petroleum and petroleum derivatives, pesticides; food, hygiene and cosmetic waste	These substances, which are thrown into the natural environment, reach water bodies by being carried by rain and surface waters
Infectious disease agents	Sewage and domestic waste	The spread of biological agents occurs by spraying into the air, infecting humans with disease-carrying animals, or by contaminating water and food with biological agents
Plant nutrients and organic fertilizers	Fertilizers applied in agricultural fields, detergents	Mixing into groundwater from agricultural areas and transport by irrigation
Pesticides	Organic and inorganic chemicals	Transportation of intensive applications in agricultural lands to water bodies
Industrial wastes (mercury, cadmium, dyes)	Solid wastes generated during or after a process in industrial and production facilities	Discharge of waste generated as a result of human activities
Transported deposits from nature	Forest fires, unconscious and intense tree cutting, soil erosion	Intense urbanization, climate changes, relocation with heavy rains
Solid waste	Metal, plastic, etc.	Discharge to soil and indirectly to the aquatic environment as a result of human activities
Petroleum products	All wastes from extraction, refining and processing processes	Oil exploration, drilling and transfer processes
Antiepileptics, antihistamines, antibiotics, antibacterial drugs, painkillers and antipyretic drugs, beta-blockers, cholesterol drugs, chemotherapy drugs, etc. are drugs detected in the ecosystem through various researches	Pharmaceutical substances have the potential to be found in wastewater and mix easily with natural waters, as they are mostly water-soluble biologically active compounds	Particularly, living creatures living in receiving environments (river, etc.) where wastewater treatment plant effluents are discharged are exposed to the most damage

(continued)

Table 7.1 (continued)

Contamination to the natural environment	Source	Pathway
It consists of substances such as endocrine disrupting chemicals, birth control pills, steroids, industrial chemicals, pesticides, agricultural pesticides (dichloro diphenyl trichloroethane, dieldrin, lindane, atrazine), surfactants found in domestic and industrial wastewater, phthalate acid esters, polyaromatic hydrocarbons, polychlorinated dioxins formed in incineration systems in solid waste landfills, biphenyls, etc	Domestic and industrial waste	Wastewaters polluted in this way are found in all environments [5]. It has been stated that estrogenic endocrine disruptors cause a high rate of androgyny in fish and therefore they are the most important type of endocrine disrupting chemical
Personal care products represent a broad group of compounds containing active and inert substances for personal care purposes. Examples of these substances are perfumes, sprays, deodorants, sunscreens, shampoos and cosmetics	Individual use	By being discharged to the soil with domestic wastes

- Preventing unconscious destruction of nature;
- Supporting natural life (Mehndiratta et al. 2013; Prodan et al. 2014; Elleuch et al. 2018; Cao et al. 2020; Maruf 2021; Tolochko et al. 2022).

7.6 Conclusion

Water resources are among the most important natural areas where environmental problems have been observed intensively from past to present. Freshwater resources, which account for 3% of the world's total water, are under great stress due to rapid population growth, urbanization, water consumption in industry and agriculture. These problems do not remain at the local level and move from local to global scale. As the direct or indirect use of water resources by humans increases, the degradation of water resources increases. Given the continuous pollution of surface water bodies by point and diffuse sources, freshwater resources in particular need to be protected and managed. All people should have the right to safe water supplies. Clean water is

Table 7.2 Industrial solid waste generation sectors and types

Industries	Generation processes	Waste types
Food and similar industries	Processing, packaging and shipping	Meats, fat, bones, offal, vegetables, fruit, nuts, shells and grains
Textile	Weaving, processing, dyeing and transport	Cloth and fiber pieces
Ready made clothing	Cutting, sewing, sizing and pressing	Fabric, yarn, metal and plastic rubber
Timber and wood products	Sawmill, wood carriers, various wood products production units	Crushed wood, sawdust, plastic, metal, paint and solvents
Paper and related unit products	Paper production, cardboard boxes	Paper and fiber waste, chemicals, paper cups, linings, ink and fasteners
Oil refineries	Asphalt and roofing materials production unit	Asphalt and tar, felt, asbestos, paper, fabric and threads
Chemical and related industries	Production units of inorganic materials such as varnish, explosives, drugs	Organic and inorganic chemicals, metal, plastic rubber, glass, oils, paints, solvents and pigments
Leather and leather products	Leather processing, leather products unit	Pieces of leather, wire yarn, dyes, oils, process and conditioning compounds
Primary metal industries	Melting, casting, iron and spraying units	Ferrous and non-ferrous metal parts, slag, sand
Stone, gravel quarries and glass industries	Production units of glass, concrete shaping and concrete coating, asbestos and various non-mineral products	Glass, cement, clay, ceramics, asbestos, stone, paper

essential for human use and healthy ecosystems. In a period when the idea that natural areas should be protected and transferred to future generations is becoming more widespread, it is very important for the future of our water resources that activities that will harm flora and fauna are carried out consciously and in a controlled manner.

Acknowledgements Many thanks to my colleagues from Fisheries Faculty who supported me in the research for their help Atatürk University for their support in my academic studies.

References

Auta HS, Emenike CU, Fauziah SH (2017) Distribution and importance of microplastics in the marine environment: a review of the sources, fate, effects, and potential solutions. Environ Int 102:165–176. https://doi.org/10.1016/j.envint.2017.02.013

Baloyi ND, Tekere M, Maphangwa KW, Masindi V (2021) Insights into the prevalence and impacts of phthalate esters in aquatic ecosystems. Front Environ Sci 9:684190. https://doi.org/10.3389/fenvs.2021.684190

Bashir I, Lone FA, Bhat RA, Mir SA, Dar ZA, Dar SA (2020) Concerns and threats of contamination on aquatic ecosystems. Bioremed Biotechnol Sustain Approach Pollut Degrad 1–26. https://doi.org/10.1007/978-3-030-35691-0_1

Beiras R (2018) Marine pollution: sources, fate and effects of pollutants in coastal ecosystems. Elsevier

Borah P, Kumar M, Devi P (2020) Types of inorganic pollutants: metals/metalloids, acids, and organic forms. In Inorganic pollutants in water. Elsevier, pp 17–31. https://doi.org/10.1016/B978-0-12-818965-8.00002-0

Brunekreef B, Holgate ST (2002) Air pollution and health. The Lancet 360(9341):1233–1242. https://doi.org/10.1016/S0140-6736(02)11274-8

Bukola D, Zaid A, Olalekan EI, Falilu A (2015) Consequences of anthropogenic activities on fish and the aquatic environment. Poultr Fisher Wildlife Sci 3(2):1–12. https://doi.org/10.4172/2375-446X.1000138

Cao Y, Chen M, Dong D, Xie S, Liu M (2020) Environmental pollutants damage airway epithelial cell cilia: implications for the prevention of obstructive lung diseases. Thoracic Cancer 11(3):505–510. https://doi.org/10.1111/1759-7714.13323

Chaudhry FN, Malik MF (2017) Factors affecting water pollution: a review. J Ecosyst Ecograph 7(1):225–231. https://doi.org/10.4172/2157-7625.100022

Chowdhary P, Raj A, Bharagava RN (2018) Environmental pollution and health hazards from distillery wastewater and treatment approaches to combat the environmental threats: a review. Chemosphere 194:229–246. https://doi.org/10.1016/j.chemosphere.2017.11.163

Culver DC, Pipan T (2014) Shallow subterranean habitats: ecology, evolution, and conservation. Oxford University Press, USA

Daly A, Zannetti P (2007) An introduction to air pollution–definitions, classifications, and history. In: Zannetti P, Al-Ajmi D, Al-Rashied S (eds) Ambient air pollution. The Arab School for Science and Technology and The EnviroComp Institute, pp 1–14

Doğan-Sağlamtimur N, Sağlamtimur B (2018) Sucul ortamlarda ötrofikasyon durumu ve senaryoları. Niğde Ömer Halisdemir Üniversitesi Mühendislik Bilimleri Dergisi 7(1):75–82. https://doi.org/10.28948/ngumuh.386307

Elleuch B, Bouhamed F, Elloussaief M, Jaghbir M (2018) Environmental sustainability and pollution prevention. Environ Sci Pollut Res 25:18223–18225. https://doi.org/10.1007/s11356-017-0619-5

Feng C, Wu F, Zhao X, Li H, Chang H (2012) Water quality criteria research and progress. Sci China Earth Sci 55:882–891. https://doi.org/10.1007/s11430-012-4384-5

Gupta A, Gupta A, Jain K, Gupta S (2018) Noise pollution and impact on children health. Indian J Pediatr 85(4):300–306. https://doi.org/10.1007/s12098-017-2579-7

Hatra G (2018) Radioactive pollution: an overview. Holist Approach Environ 8(2):48–65. https://hrcak.srce.hr/202085

Kamtchueng BT, Fantong WY, Ueda A, Tiodjio ER, Anazawa K, Wirmvem MJ, Hell JV (2014) Assessment of shallow groundwater in Lake Nyos catchment (Cameroon, Central-Africa): implications for hydrogeochemical controls and uses. Environ Earth Sci 72:3663–3678. https://doi.org/10.1007/s12665-014-3278-6

Karri RR, Ravindran G, Dehghani MH (2021) Wastewater—sources, toxicity, and their consequences to human health. In: Soft computing techniques in solid waste and wastewater management. Elsevier, pp 3–33. https://doi.org/10.1016/B978-0-12-824463-0.00001-X

Kazour M, Terki S, Rabhi K, Jemaa S, Khalaf G, Amara R (2019) Sources of microplastics pollution in the marine environment: Importance of wastewater treatment plant and coastal landfill. Mar Pollut Bull 146:608–618. https://doi.org/10.1016/j.marpolbul.2019.06.066

Khayat S, Marei A, Natsheh B, Abu-Khalaf N (2012) Mechanisms of groundwater pollutants transport in Tulkarm area/Palestine. Resour Environ 2(6):281–290. https://doi.org/10.5923/j.re.20120206.06

Landrigan PJ, Stegeman JJ, Fleming LE, Allemand D, Anderson DM, Backer LC, Rampal P (2020) Human health and ocean pollution. Ann Glob Health 86(1). https://doi.org/10.5334/aogh.2831

Lin L, Yang H, Xu X (2022) Effects of water pollution on human health and disease heterogeneity: a review. Front Environ Sci. https://doi.org/10.3389/fenvs.2022.880246

Malik DS, Sharma AK, Sharma AK, Thakur R, Sharma M (2020) A review on impact of water pollution on freshwater fish species and their aquatic environment. Adv Environ Pollut Manage Wastewater Impacts Treatment Technol 1:10–28. https://doi.org/10.26832/aesa-2020-aepm-02

Maruf A (2021) Legal aspects of environment in Indonesia: an efforts to prevent environmental damage and pollution. J Hum Rights Cult Legal Syst 1(1). https://doi.org/10.53955/jhcls.v1i1.4

Masindi V, Muedi KL (2018) Environmental contamination by heavy metals. Heavy Metals 10:115–132

Mateo-Sagasta J, Zadeh SM, Turral H, Burke J (2017) Water pollution from agriculture: a global review. Executive summary

Mehndiratta P, Jain A, Srivastava S, Gupta N (2013) Environmental pollution and nanotechnology. Environ Pollut 2(2):49. https://doi.org/10.5539/ep.v2n2p49

Mostofa KM, Liu CQ, Zhai W, Minella M, Vione D, Gao K, Sakugawa H (2016) Reviews and syntheses: ocean acidification and its potential impacts on marine ecosystems. Biogeosciences 13(6):1767–1786. https://doi.org/10.5194/bg-13-1767-2016

NSTC (2003) An assessment of coastal hypoxia and eutrophication in U.S. waters. National Science and Technology Council, Committee on Environment and Natural Resources

Patel N, Khan MD, Shahane S, Rai D, Chauhan D, Kant C, Chaudhary VK (2020) Emerging pollutants in aquatic environment: source, effect, and challenges in biomonitoring and bioremediation-a review. Pollution 6(1):99–113. https://doi.org/10.22059/poll.2019.285116.646

Prodan M, Ghicioi E, Oancea D (2014) Correlation of explosion parameters and explosion type events for preventing environmental pollution. Environ Eng Manag J 13(6):1409–1414

Rajaram T, Das A (2008) Water pollution by industrial effluents in India: discharge scenarios and case for participatory ecosystem specific local regulation. Futures 40(1):56–69. https://doi.org/10.1016/j.futures.2007.06.002

Saha JK, Selladurai R, Coumar MV, Dotaniya ML, Kundu S, Patra AK (2017) Soil pollution-an emerging threat to agriculture. Springer Singapore, pp 271–315. https://doi.org/10.1007/978-981-10-4274-4

Schindler DW (2012) The dilemma of controlling cultural eutrophication of lakes. Proc Royal Soc B: Biol Sci 279(1746):4322–4333. https://doi.org/10.1098/rspb.2012.1032

Schweitzer L, Noblet J (2018) Water contamination and pollution. In Green chemistry. Elsevier, pp 261–290. https://doi.org/10.1016/B978-0-12-809270-5.00011-X

Sharma BM, Melymuk L, Bharat GK, Přibylová P, Sáňka O, Klánová J, Nizzetto L (2018) Spatial gradients of polycyclic aromatic hydrocarbons (PAHs) in air, atmospheric deposition, and surface water of the Ganges River basin. Sci Total Environ 627:1495–1504. https://doi.org/10.1016/j.scitotenv.2018.01.262

Sonone SS, Jadhav S, Sankhla MS, Kumar R (2020) Water contamination by heavy metals and their toxic effect on aquaculture and human health through food Chain. Lett Appl NanoBioScience 10(2):2148–2166. https://doi.org/10.33263/LIANBS102.21482166

Tiwary A, Williams I (2018) Air pollution: measurement, modelling and mitigation. CRC Press

Tolochko S, Bordiug N, Les T (2022) Features of formation of environmental competence of education seekers in the context of preventing environmental pollution during war. SciRise: Pedagogical Educ 3(48):4–10. https://doi.org/10.15587/2519-4984.2022.257021

Ukaogo PO, Ewuzie U, Onwuka CV (2020) Environmental pollution: causes, effects, and the remedies. In Microorganisms for sustainable environment and health. Elsevier, pp 419–429. https://doi.org/10.1016/B978-0-12-819001-2.00021-8

Vo QT, Künzer C, Vo QM, Moder F, Oppelt N (2012) Review of valuation methods for mangrove ecosystem services. Ecol Ind 23:431–446. https://doi.org/10.1016/j.ecolind.2012.04.022

Zhou XY, Lei K, Meng W, Khu ST (2017) Industrial structural upgrading and spatial optimization based on water environment carrying capacity. J Clean Prod 165:1462–1472. https://doi.org/10.1016/j.jclepro.2017.07.246

Chapter 8
Toxicology Mechanisms of Pollutants

Gonca Alak

Abstract Contaminants can have toxic effects to aquatic organisms. This study aims to bring together the toxicity mechanisms of contaminants that somehow affect aquatic organisms and the reference parameters that play a role in the effectiveness of these mechanisms. Since the contaminants (pesticides, plastics, global warming, bacteria, or heavy metals) found in the aquatic ecosystem are thought to pose a threat to aquatic organisms, there is still a need to develop a common vocabulary for these contaminants and to thoroughly research their toxicity mechanisms.

Keywords Aquatic toxicology · Toxicology · Fish · Toxicity mechanism

List of Abbreviations

8-OHdG	8-Hydroxy-2′-deoxyguanosine
Ach	Acetylcholine
AChE	Acetylcholinesterase
AOP	Adverse outcome pathway
AOPP	Enhanced oxidative protein products
ARE	Antioxidant response element
Bcl-2	B-cell lymphoma 2
BDNF	Brain-derived neurotrophic factor
BRI	Biomarker response index
CAT	Catalase
COX-2	Cyclooxygenase-2
DNA	Deoxyribonucleic acid
GPx	Glutathione peroxidase

G. Alak (✉)
Department of Seafood Processing Technology, Faculty of Fisheries, Atatürk University, Erzurum, Türkiye
e-mail: galak@atauni.edu.tr

M. Atamanalp et al., *Aquatic Toxicology in Freshwater*, Springer Water,
https://doi.org/10.1007/978-3-031-56669-1_8

GR	Glutathione reductase
GSH	Glutathione
GST	Glutathione-S-transferase
HO-1	Heme oxygenase-1
IBR	Integrated biomarker response
IL	Interleukins
iNOS	Inducible nitric oxide synthase
Keap1	Kelch like-ECH-associated protein 1
KERs	Key event relationships
Kes	Key events
MDA	Malonaldehyde
MIE	Molecular initiation event
NAD(P)H	Nikotinamid adenin dinükleotit fosfat
NF-κB	Nuclear factor kappa B
NGF	Nerve growth factor
NO	Nitric oxide
NQO1	Quinone oxidoreductase 1
Nrf-2	Nuclear factor erythroid-2
NT	Neurotrophin
p75NTR	Pan-neurotrophic receptor
ROS	Reactive oxygen species
SOD	Superoxide dismutase
TNF	Tumor necrosis factor
TNF-α	Tumor necrosis factor alpha
Trk	Tyrosine kinase receptor

8.1 Introduction

The community, including bacteria, phytoplankton, higher aquatic plants, invertebrates, vertebrates, and other living organisms, is involved in a variety of biological processes (Ogamba et al. 2023). Compared to conventional animal models of toxicity, aquatic invertebrates have a significant position in terms of organism complexity. Especially in recent years, more aquatic organisms have been used in experimental research. Aquatic species are excellent candidates for ecotoxicology because of their capacity to absorb and store pollutants in a variety of ways, including through the digestive system and gills when ingesting food or water (Arslan and Günal 2023).

8.1.1 Pollutants as Stressors

In aquatic ecosystems, which are very important for a sustainable environment (Kumar et al. 2023; Özgeriş et al. 2023; Alak et al. 2023; Atamanalp et al. 2023; Uçar et al. 2023), as in all vertebrates, the general physiological response to threatening situations in fish is called 'stress'. The organism's stress response kicks in soon as it perceives a stressor. The physiological response triggered by stressors (stress response) is a set of adjustments known as internal homeostasis that can maintain a certain level and maintain vitality (Petitjean et al. 2019). Stress response is an adaptive response of organisms sent to a stimulus (stressor) perceived as dangerous that helps maintain homeostasis (Uçar et al. 2021). These responses are grouped into primary, secondary and tertiary phases (Atamanalp et al. 2022). The primary phase refers to a generalized neuroendocrine response. Catecholamines (epinephrine and norepinephrine) and cortisol are released from chromaffin and interrenal cells, respectively. Higher circulating levels of these hormones trigger a secondary response. The secondary phase presents as hyperglycemia resulting from enhanced glycogenolysis and gluconeogenesis, vasodilation of arteries in the gill filaments, increased cardiac output and depression of immune function. The first two phases are considered adaptive and maintain homeostasis by enabling fish to adapt to stress factors. In contrast, the tertiary phase includes systemic changes in which organisms are unable to adapt to stress factors, leading to adverse effects on this situation. This includes the impact on behavior, wellbeing, development, and reproduction, as well as swimming ability (Barton 2002; Patel et al. 2022) (Fig. 8.1). Toxicity tests are widely used to assess the potential hazard of contaminants to aquatic organisms (Alak et al. 2017; Uçar et al. 2021; Kumar et al. 2023; Özgeriş et al. 2023). Referring to the above mechanisms, when fish are exposed to stress factors, different processes have an effect on different tissue levels (cells, tissues, organs, whole organism). For example, since they live in an aquatic environment, fish, which are ectothermic creatures, are frequently impacted by daily or seasonal temperature fluctuations. During acute heat stress, the fish's blood permeability and other blood variables change (Bard et al. 2019) and the fish's immune system is also suppressed. Together, they regulate secondary stress response factors that alter the distribution of essential resources such as energy and oxygen to tissues and organs, and also compromise hydro-mineral imbalance and the immune system. However, it is known that long-term exposure to stress negatively affects basic life functions (growth, development, disease resistance, energy balance, behavior and reproduction) (Fig. 8.1) (Schreck and Tort 2016; Li et al. 2022).

8.1.2 Physiological Effects of Pollutants

In the evaluation of fish health, changes in blood parameters associated with acute and/or chronic environmental contaminants have drawn increasing attention

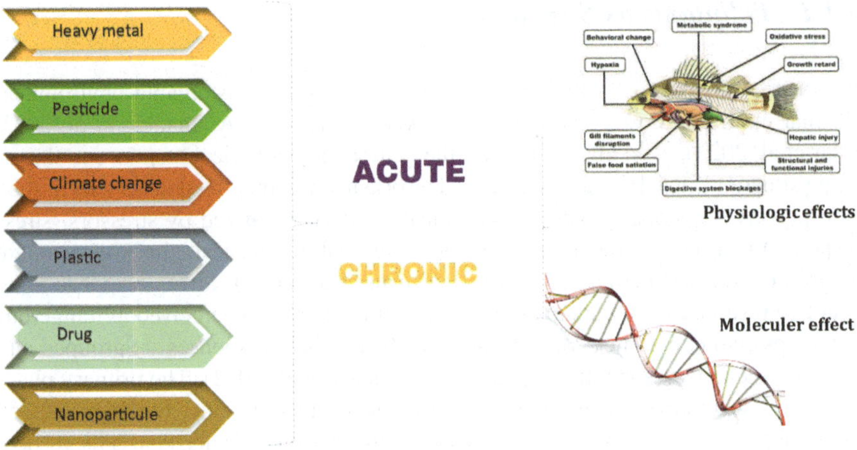

Fig. 8.1 Potential effects of different pollutants on aquatic organisms

(Corredor-Santamaría et al. 2016). Variations in hematocrit, hemoglobin concentration, leukocyte and erythrocyte counts from hematological parameters have been used as physiological indicators of pollution and organic dysfunction in both environmental and aquaculture studies. These parameters are also widely applied as prognostic and diagnostic tools in fish health condition (Burgos-Aceves et al. 2019). As the measurement of hematology indices provides comprehensive information, it is used in fisheries practices and research on the effects of water contamination on fish organisms to evaluate fish health criteria such as the physiological state of the organism, oxygen carrying capacity, immune functions, stress and various metabolic functions (Fazio 2019; Bojarski and Witeska 2020) (Fig. 8.2).

A. Oxidative stress markers

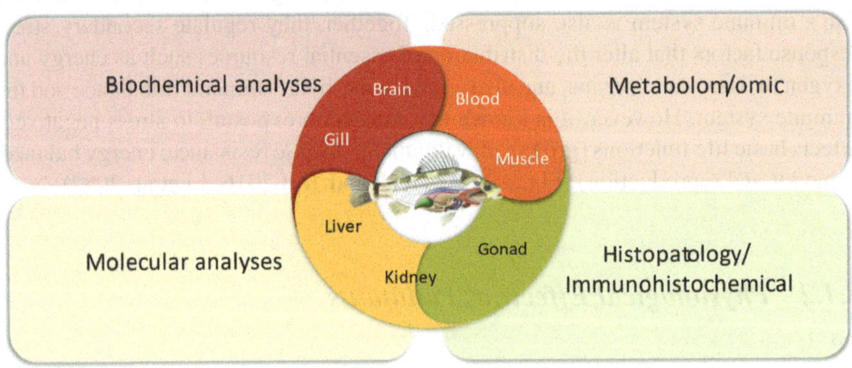

Fig. 8.2 Target tissue and commonly used assays of pollutant exposure

It is advantageous to strategically use antioxidants to determine the effects of contaminants on organisms, to understand the causes and control of oxidative stress (Arslan and Günal 2023). Antioxidants can be classified into four categories according to the role they play in the defense system activated against oxidative stress. The first line of defense consists of antioxidants that inhibit oxidative processes by suppressing the formation of radical species. These antioxidants act very quickly and neutralize any free radical that can induce the production of other radicals or any type of chemical that has the potential to transform into a free radical. They degrade the superoxide anion radical, decomposing hydrogen peroxide and hydroperoxides giving rise to harmless species (e.g. molecular oxygen). The second line of defense is formed by antioxidants that suppress chain initiation or break chain propagation reactions (radical scavenging antioxidants). This scavenging ability is accomplished by electron donation. Newly formed radicals can be easily neutralized and become non-damaging species. This category includes hydrophilic (ascorbic acid, glutathione, uric acid) and lipophilic (alpha tocopherol and ubiquinol) antioxidants. The third line of defense includes repair (also called "de novo") antioxidants, which act after free radical damage has occurred. This group consists of enzymes that repair damaged deoxyribonucleic acid (DNA), lipids and proteins. These antioxidants can recognize, degrade, remove oxidized/damaged proteins, lipids and DNA and inhibit their cellular accumulation. The most common examples of this type are DNA repair enzymes (polymerases, glycosylases, and nucleases), proteolytic enzymes (peptidases, proteases, and proteinases), which are found in both the mitochondria and cytosol of mammalian cells. The fourth line of defense includes antioxidants that act by exploiting adaptation mechanisms: They prevent the formation of free radicals or reactions involving radical species by using the signals necessary for the generation of free radicals or for reactions involving free radicals. In the adaptation mechanism, the signal produced by a free radical species induces the synthesis and transport of an adequate antioxidant to the appropriate site (Alak et al. 2017, 2018; Ighodaro and Akinloye 2018).

Xentebiotics directly affect the physiological process and metabolic demand of fish (Baldissera et al. 2019; Atamanalp et al. 2022; Köktürk et al. 2022a, b, c, 2023; Alak et al. 2023; Uçar et al. 2023; Özgeriş et al. 2023). Although fish can cope with a range of changes through physiological plasticity or evolution, severe illness and death occur when tolerances are exceeded. Many studies have revealed that contaminants can induce oxidative stress in fish and activate the enzyme-induced and non-enzyme-induced antioxidant defense system (Alak et al. 2017; Gutteridge et al. 2018). In order to maximally prevent oxidative damage caused by excessive production of reactive oxygen species (ROS), the body can engage the antioxidant defense system as a survival strategy by eliminating excessive ROS production caused by contaminant stress. These include low molecular weight antioxidants such as glutathione (GSH) and high molecular weight defenses with enzymes such as superoxide dismutase (SOD), catalase (CAT) as the first line of defense, enzymes related to glutathione metabolism such as glutathione peroxidase (GPx) [glutathione reductase (GR)] and Glutathione-S-transferase (GST) and glutathione reductase (GR)] (Atamanalp et al. 2021, 2023; Uçar et al. 2022). In both in vitro and in vivo studies, it has been reported that contaminants can induce oxidative stress by triggering the formation of ROS. Stimulation of oxidative stress causes extensive damage to biomolecules such as DNA, proteins, lipids and cellular membranes (Atamanalp et al. 2021, 2023). Oxidative stress in the cellular system is produced as a result of an imbalance between oxidative and reducing processes and is typically induced when the physiological antioxidant defense system is unable to counteract the increased ROS level (Koner et al. 2021).

Oxidative stress is involved in different normal cellular functions by acting as a second messenger in signal transduction, but can also cause cellular damage, including oxidative damage to proteins and DNA as well as lipid peroxidation. Environmental stress can lead to a state of oxidative stress in aquatic organisms. For example, it has previously been observed that thermal stress can modulate the oxidative stress state in various tissues of teleost fish (Dettleff et al. 2022a, b). It has been reported that thermal stress at low temperature can inhibit antioxidant enzymes and cause oxidative damage in dairy fish (*Chanos chanos*) (Chang et al. 2022), similarly, lipid peroxidation was observed in Pacu (*Piaractus mesopotamicus*) under the same stressor effect (Bacchetta et al. 2020; Ale et al. 2021). It is also known that lipid peroxidation may vary according to species, carnivorous species have very low GPx activity in liver mitochondria compared to herbivorous and omnivorous species, and have high levels of SOD and CAT activity in their livers, and these differences may be effective in enzymatic antioxidant system activities (Parlak et al. 2022; Yeltekin et al. 2022). Starting from this point, different amounts of ROS are produced in cells. The ROS produced reflects the function of biotransformation mechanisms of aquatic invertebrates. Low concentrations of ROS express physiological properties, including immune response, while high concentrations of ROS originate from the biomolecules of cells such as protein and lipids and alter their function. Due to the negative effects of these biomolecules on cells and tissues, it becomes necessary to remove ROS from cells, regulate homeostasis within the cell, and prevent pathological development in tissues. In this way, antioxidant enzymatic and non-enzymatic systems in cells keep ROS at normal concentrations. Glutathione is one of these non-enzymatic antioxidant systems to maintain cellular homeostasis, while enhanced oxidative protein products (AOPP) are a novel oxidative stress marker of non-enzymatic systems.

2 Cytokines markers

Similar to the screening of oxidative stress parameters, cytokines are multifunctional, soluble proteins that provide communication between cells and play a key role in cellular development—growth, tissue repair, homeostasis, tumorigenesis, immunity and hemopoiesis. These proteins are released by cells, affecting the functions of other cells (Fig. 8.3). There are many types of cytokines that generally affect hormones and growth factors, such as chemokines, interleukins (IL), tumor necrosis factor (TNF), lymphokines, interferons (Uçar et al. 2022; Alak et al. 2023). Basic proinflammatory cytokines are composed of tumor necrosis factor alpha (TNF-α) and IL-6. Pro-inflammatory cytokines are molecules involved in initiating and regulating

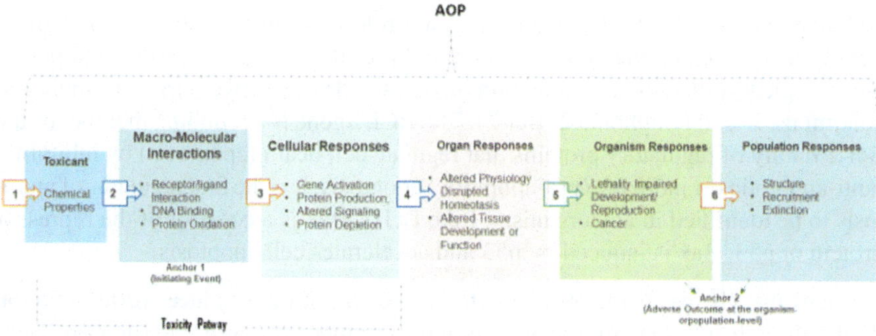

Fig. 8.3 Potential pathway diagram of toxicity exposure

fish innate immune responses, and their induction are primary indicators of inflammatory processes. TNF-α is a pleiotropic cytokine that regulates many physiological events, pathological processes including cell survival, apoptosis and necrosis, the development of chronic inflammatory diseases, and plays a role in cell growth, proliferation, apoptosis, inflammation and immune mechanisms (Sakai et al. 2021; Brix et al. 2022). TNF-α genes have been identified in various fish species, but studies on their biological functions are continuing. They can promote the secretion and release of other cytokines, activate neutrophils and lymphocytes, increase the permeability of endothelial cells, control cellular metabolic activities, and participate in brain injuries (Farag et al. 2022). All these cytokines are produced by immune cells such as B lymphocytes, T lymphocytes and mast cells through tissue injury and antibody production. The earliest cytokines synthesized and released are TNF-α and IL-1. TNF-α is produced by monocyte macrophages (liver kuppfer cells) and lymphocytes. Its target cells are endothelial cells, monocytes, macrophages, neutrophils, and fibroblasts, which stimulate the production and release of IL-1, IL-6, IL-8, and TNF-α in these cells.

TNF-α, a transmembrane protein identified for the first time to be secreted from fat α-macrophages, is a cytokine with important pro-inflammatory properties that modulates immune functions (innate and adaptive immunity), plays a role in cell proliferation and apoptosis. This cytokine is synthesized by many cells such as T cells, smooth muscles, fibroblasts, endothelial cells, adipocytes, B cells, especially macrophages and monocytes (Uçar et al. 2022). TNF-α has a broad spectrum of action on neurons that can be both neuroprotective and neurotoxic (Alak et al. 2023). It is the most potent pro-inflammatory cytokine produced in response to tissue injury or chronic inflammation (Sakai et al. 2021). One of the most important factors in TNF-α signal transduction is regulated by the transcription factor, nuclear factor kappa B (NF-κB) (Yeltekin et al. 2022) and inhibition of NF-κB contributes to the suppression of tumor cell proliferation, induction of apoptosis, and inhibition of invasion. NF-κB also controls the production of inducible nitric oxide synthase (iNOS) and COX-2 in the mechanism of inflammation (Altınöz and Elmacı 2018). iNOS is one of the three key enzymes that produce nitric oxide (NO) from the amino acid l-arginine.

NO, derived from iNOS, plays an important role in numerous physiological (e.g., blood pressure regulation, wound repair, and host defense mechanisms) and pathophysiological (inflammation, infection, neoplastic diseases, liver cirrhosis, diabetes) conditions. B-cell lymphoma 2 (Bcl-2): The BCL2 gene is a founding member of the Bcl-2 family of regulatory proteins that regulate cell death (apoptosis) by inhibiting (anti-apoptotic) or inducing (pro-apoptotic) apoptosis. It is the first regulator of apoptosis to be identified in any organism. Like Bcl-2, Bax is a cofactor of the repressor protein of p53. Bax is induced by p53 and accelerates cell apoptosis.

Interleukin-6 (IL-6): A pleiotropic cytokine, IL-6 not only induces differentiation of B cells into plasma cells in the humoral immune response, but also produces antibodies for antibody-secreting cells (Uçar et al. 2022). It is a pleotropic cytokine with broad-spectrum biological activity that plays a role in the regulation of inflammation, cell differentiation, cell proliferation, regulation of the immune system, hematopoiesis and tumorigenesis. It is also important because it is effective in stimulating the production of immunoglobulin from B cells, T cell activation, stimulation of IL-2 synthesis, cell cycle maintenance and apoptosis inhibition (Uçar et al. 2021, 2022, 2023). Although its release is strictly controlled by transcriptional and post-transcriptional mechanisms, the unregulated and sustained synthesis of IL-6 exerts a pathological effect on chronic inflammation and autoimmunity. The cytokine breaks down pro-interleukin-1β (pro-IL-1β) and pro-IL-18 into their mature and biologically active forms. Increased protein levels of pro-IL-1β and IL-1β have been proven to accompany NF-kB pathway activation (Nan et al. 2021).

Nuclear factor erythroid-2 (Nrf-2): Antioxidant defense mechanisms are of great importance for cells that are constantly exposed to free radicals throughout normal life. Cells use enzyme systems called Phase I and Phase II in these mechanisms. Nerve cells mainly protect themselves using Phase II and antioxidant enzymes (GSH, SOD, CAT, GR GST and GPx). Nrf2 is a transcription factor required for protection against oxidative damage, regulating the expression of antioxidant and phase II detoxification enzymes. Nrf2/Keap1 pathway, which is accepted as the main regulator of cellular defense mechanisms against cellular stress, provides survival of the cell under stress conditions by regulating the gene expressions of various cell protective protein networks involved in the repair or elimination of damaged macromolecules (Yuan et al. 2018). Nrf2 is highly expressed as a regulator, especially in tissues where detoxification reactions take place. A. In normal conditions: Nrf2 is retained in the cytoplasm by Kelch like-ECH-associated protein 1 (Keap1). B. When oxidative stress increases: Nrf2 cleaves from the Keap1 repressor and translocates to the nucleus to form a heterodimer with related sMaf proteins. By binding to ARE, this heterodimer increases the expression of antioxidant enzymes such as NAD(P)H, quinone oxidoreductase (NQO1), GST, heme oxygenase-1 (HO-1) and GPx, which have important roles in cellular defense (Keap1: Kelch like-ECH-associated protein 1; ARE: antioxidant response element) Nrf2 is a key transcriptional activator of genes responsible for maintaining redox homeostasis through synthesis/recycling of reduced glutathione (GSH), an essential brain antioxidant (Spagnuolo et al. 2020) and plays a very important role in the increase of ROS from the metabolism of environmental contaminants and in the process of coping with oxidative stress. Neurotoxicity is an etiological phenomenon of a change in the structure and function of the central nervous systems by a biological, chemical or physical agent (Uçar et al. 2022; Yeltekin et al. 2022). An organism's homeostasis depends on bidirectional communication between the neuroendocrine system and the immune system to provide adaptive responses to physiological disturbances and pathogenic challenges. Although it was initially thought that the neuroendocrine system and the immune system acted independently, it is accepted in recent publications that a widespread communication network controls a harmonious neuroendocrine-immune interaction. Given the dysregulation of redox/oxidative balance and apoptotic signaling in neuronal apoptosis induced by oxidative stress, mitochondrial dysfunction may be a primary contributor to brain aging and neuronal pathologies (Khairy and Attia 2021). Again, Nrf2-deficient neurons are highly susceptible to oxidative stress and are recovered by transfecting with a functional Nrf2 construct (Alak et al. 2023).

3 DNA damage markers

During the formation of free radicals, oxidative damage occurs in DNA first and then the production of 8-hydroxy-2′-deoxyguanosine (8-OHdG) causes the release of significant amounts of DNA oxidation products. The generation of ROS has a significant effect on causing damage to various cell membrane components, ranging from antioxidant enzyme inhibitions to the regulation of malonaldehyde (MDA), redox-sensitive signal transduction pathways and cell death (apoptosis) (Alak et al. 2018; Uçar et al. 2021).

Oxidative stress induced by contaminants can damage cellular macromolecules such as lipids, DNA and proteins. Cell death, lipid peroxidation, DNA damage and protein peroxidation occur by apoptotic or necrotic mechanisms following oxidative stress (Wang et al. 2016). It is known that apoptosis (programmed cell death) is a vital phenomenon and therefore many disorders and diseases can cause cell proliferation/differentiation imbalance and cell death in fish. 8-OHdG is the most common and sensitive of the twenty-three oxidative base damage caused by ROS in DNA. The DNA nucleotides are common targets for oxidative damage due to their significant mutagenic and oxidative capacity. Guanine is most prone to oxidation of all nucleic base pairs. When a free hydroxyl group oxidized guanine, 8-OHdG is formed, a common form of DNA degradation caused by free radicals. Levels of 8-OHdG have been studied in aquatic organisms and have been recognized as a biomarker of oxidative stress caused by contamination (El-Agri et al. 2022). 8-OHdG, which is the most important marker of oxidative damage and occurs as a result of stimulation of DNA, increases due to oxidative stress. Apoptosis occurs in two different ways: receptor mediated apoptotic pathway and mitochondrial mediated apoptotic pathway; both work with caspase-3 as effector caspase (Uçar et al. 2021). Caspase-3 is an important molecule in the regulation of both mitochondrial and apoptotic pathways (Atamanalp et al. 2021), It can be activated by ROS production and therefore induce cell apoptosis and necrosis (Uçar et al. 2021).

As another marker of brain function, acetylcholinesterase (AChE) activity, which plays an important role only for cholinergic signaling, is one of the markers investigated, due to hydrolysis of acetylcholine and thus ending the synaptic effect of acetylcholine (Spagnuolo et al. 2020). AChE is a key enzyme of the neurotransmitter that hydrolyzes the neurotransmitter acetylcholine (ACh), regulates synaptic cleft levels (Serafini et al. 2019), can reduce acetylcholine in biological nerve transmission, synaptic space, and terminate the excitatory effect of neurotransmitter on the postsynaptic membrane, thus ensuring the normal transmission of nerve signals in the organism (Zhang et al. 2017). ACh-mediated neurotransmission is one of the most important functions of the nervous system, and this molecule serves as a preganglionic transmitter in the sympathetic nervous system and a post-ganglionic transmitter in the parasympathetic nervous system, as well as an excitatory function (Olivares-Rubio and Espinosa-Aguirre 2021). Cholinesterases have long been recognized as remarkable biomarkers and biosensors in monitoring environmental contamination (Umar and Aisami 2020). It is a protein of the neurotrophin family that contains brain-derived neurotrophic factor (BDNF), neurotrophin (NT), nerve growth factor (NGF) and has important physiological roles such as synaptic plasticity, neurogenesis and neuronal maturation (Topal et al. 2021).

Neurotrophins are a family of polypeptide growth factors that affect the viability and function of neurons and control synaptic function and synaptic plasticity. Neurotrophin family members are nerve growth factor, BDNF, neurotrophin-3, neurotrophin-4/5, neurotrophin-6, and neurotrophin-7. Neurotrophins are synthesized from many cell types in the central nervous system, peripheral nervous system neurons, and peripheral tissues. Neurotrophins have important functions such as supporting the vitality of neurons, mediating immune responses, influencing glucose and cholesterol metabolism, and stimulating lipid oxidation in muscle cells. The BDNF, a highly active neurotrophic factor and a member of the neurotrophin growth factor family, plays a role in the development, vitality and maintenance of neurons, and decreased stress-related neurogenesis (Khairy and Attia 2021). The main function of BDNF is to ensure the functionality of hippocampal/cortical neurons, cholinergic neurons, and peripheral sensory neurons. BDNF is mainly responsible for stimulating neuroplasticity and adult neurogenesis by stimulating the growth and differentiation of neurons, synapse assembly, and supporting the survival of existing neurons (Lesniak et al. 2021). BDNF is involved in many stages in the development of brain tissue and in the process of neuronal development, such as neuronal migration, neuronal survival/protection, neuronal excitation, induction of neurotransmitter and neuropeptide synthesis. It has been determined that BDNF is synthesized from non-neuron cells, peripheral vascular endothelial cells, lymphocytes, platelets, leukocytes, monocytes, T and B cells (Murer et al. 2001). It has been reported that BDNF also has functions in the immune system and stimulates lipid oxidation in muscle cells (Uçar et al. 2022). BDNF gene codes have been found in fish, amphibians, reptiles and mammals (Murer et al. 2001). Neurotrophins exert their effects through the tyrosine kinase receptor (Trk) and pan-neurotrophic receptor (p75NTR) to which they tend to bind. p75NTR, the first isolated neurotrophin receptor, is within the TNF receptor family, and has a glycoprotein structure. There are three types of specifically acting tyrosine kinase receptor, TrkA, TrkB, and TrkC. p75NTR signaling, Trk receptors mediate the biological activities of nerve cells related to life and development, and specifically the programming, initiation and execution of apoptosis (Manti et al. 2023).

8.1.3 Approach to Integrated Biomarker Response

When exposed to stress related to pollutants, the organism undergoes molecular, cellular and physiological changes (Atamanalp et al. 2022, 2023). Aquatic toxicity assays are considered important methods in detecting these changes (Uçar et al. 2021). A single biomarker may not be effective in a true and practical assessment of the toxicity of the toxicant to aquatic life forms. Therefore, it is recommended to use a combined biomarker approach to understand the organism's response to toxic substances (Uçar et al. 2021). Integrated Biomarker Response (IBR) provides a comprehensive understanding that combines all biomarker responses into a single cluster (Serafim et al. 2012). The Biomarker Response Index (BRI) has been widely used in recent years to integrate multiple biomarker responses and mainly focuses on the level of change of biomarker responses to determine the health status of the test organism under toxic stress (Bej et al. 2021). A range of fish biological responses may be an early indicator of contamination and damage to the organism. These biological responses, called biomarkers, follow a path from the molecular level to the organism, including molecular, biochemical, physiological and structural changes (de Moraes et al. 2018). Biomarkers are frequently used in ecotoxicology as they allow to study the toxic effects that occur at low concentrations of exposure. However, most toxicity studies evaluate cellular biomarkers that lack ecological relevance. Proper selection of biomarkers is a key process for effectively linking subindividual responses to negative individual outcomes. This selection should take into account the mode of action of the toxicant, the physiological mechanisms and pathways potentially affected, as well as the specificity and sensitivity of the response to the toxicant. A biosensor is defined as a contaminant-induced change in the biochemical or cellular components of a process, structure, or function that can be a sensor in biological systems. A biomarker is an indicator in a living organism that reflects non-fatal molecular and/or cellular changes (biochemical, physiological and histological) that occur along a metabolic pathway as a result of exposure to toxic or physiological changes (Mehrotra 2016).

An adverse outcome pathway (AOP) is a conceptual model characterizing the link between a stressor-induced molecular initiation event (MIE) and consequent key events (KEs) through increasingly higher levels of biological organization and ultimately connecting the MIE through a set of key event relationships (KERs). Adverse outcome pathways are generally not stressor-specific but rather describe a physiological sequence of events triggered by a particular MIE that can be induced by more than one stressor. Adverse outcome pathways provide a relatively standardized model framework for integrating a wide range of information on how stress factors elicit responses at various levels of biological organization in an organism. Additionally, AOPs are a useful tool in uncovering important uncertainties in our understanding of how a stressor can have effects on organisms (Brix et al. 2022).

In addition to all these mechanisms of action, there is a phenomenon called hormesis. This phenomenon is also characterized by a utility effect on the response to low doses of the toxic substance, but inhibitory at high doses, but the word utility should be used with caution. The phenomenon of hormesis has been widely studied in various fields in recent years. Hormesis occurs, for example, in the application of small amounts of both biotic and non-biotic stress factors in the study of the positive effect of mild stress on animal performance. When examining the behavior of a toxic substance in the chosen environment, there are a number of relevant properties defined through utility functions. Detection and characterization of the hormetic effect in toxicological tests is also defined by such a utility function (Pozuelo-Campos et al. 2023).

In addition to being involved in the pathogenesis of localized tissue or organ cytotoxic reactions, chemically reactive metabolites also play an important role in mediating carcinogenic and teratogenic adverse reactions as well as adverse drug reactions characterized by systemic hypersensitivity manifestations.

8.2 Conclusion

It is known that all kinds of xentobionts can cause malformations in aquatic organisms, toxicity in tissues belonging to systems such as blood and gill tissues as well as reproductive-digestive system, and even cause death.

Oxidative stress, generation of reactive oxygen species in response to cell damage leads to tissue/organ/system-based toxicity. Increased oxidative stress and lack of antioxidant factor play an important role in contaminant-induced toxicity. Xenobiotics increase the production of reactive oxygen species, impair mitochondrial function and cause mutations in mitochondrial DNA. The presence of tissue damage due to mitochondrial damage and oxidative phosphorylation and the decrease in the amount of antioxidant defense (catalase, superoxide dismutase and peroxidase) creates a picture of eutrophy that leads to apoptosis.

In the light of the information summarized above, it is thought that regardless of origin, contaminants trigger oxidative stress and oxidative lipid damage in aquatic organisms and that aquatic ecosystem health, especially water contamination, is important in these organisms. It can be concluded that aquatic pollution can be prevented, free radical reactions and oxidative lipid damage in organisms can be reduced by actions such as water footprint and monitoring of water quality criteria to be applied regularly, thus helping to reduce toxicity through oxidative stress mechanism and prevent oxidative lipid damage.

Acknowledgements Many thanks to my colleagues from Fisheries Faculty who supported me in the research for their help Atatürk University for their support in my academic studies.

References

Alak G, Yeltekin AÇ, Tas IH, Ucar A, Parlak V, Topal A, Atamanalp M (2017) Investigation of 8-OHdG, CYP1A, HSP70 and transcriptional analyses of antioxidant defence system in liver tissues of rainbow trout exposed to eprinomectin. Fish Shellfish Immunol 65:136–144. https://doi.org/10.1016/j.fsi.2017.04.004

Alak G, Ucar A, Yeltekin AÇ, Çomaklı S, Parlak V, Taş IH, Türkez H (2018) Neuroprotective effects of dietary borax in the brain tissue of rainbow trout (*Oncorhynchus mykiss*) exposed to copper-induced toxicity. Fish Physiol Biochem 44:1409–1420. https://doi.org/10.1007/s10695-018-0530-0

Alak G, Yeltekin AÇ, Köktürk M, Nas MS, Parlak V, Calimli MH, Atamanalp M (2023) May PdCu@ f-MWCNT NPs be an ecotoxicologic risk? Appl Organomet Chem 37(3):e7013. https://doi.org/10.1002/aoc.7013

Ale A, Bacchetta C, Rossi AS, Scarabotti PA, Cazenave J (2021) Low temperature stress in a cultured fish (*Piaractus mesopotamicus*) fed with *Pyropia columbina* red seaweed-supplemented diet. Fish Physiol Biochem 47:829–839. https://doi.org/10.1007/s10695-021-00944-7

Altinoz MA, Elmaci İ (2018) Targeting nitric oxide and NMDA receptor-associated pathways in treatment of high grade glial tumors. Hypotheses for nitro-memantine and nitrones. Nitric Oxide 79:68–83. https://doi.org/10.1016/j.niox.2017.10.001

Arslan P, Günal AÇ (2023) Does fipronil affect on aquatic organisms? Physiological, biochemical, and histopathological alterations of non-target freshwater mussel species. Water 15(2):334. https://doi.org/10.3390/w15020334

Atamanalp M, Parlak V, Özgeriş FB, Çilingir Yeltekin A, Ucar A, Keleş MS, Alak G (2021) Treatment of oxidative stress, apoptosis, and DNA injury with N-acetylcysteine at simulative pesticide toxicity in fish. Toxicol Mech Methods 31(3):224–234. https://doi.org/10.1080/15376516.2021.1871794

Atamanalp M, Ucar A, Alak G (2022) Catfish as an ecotoxicological model for assessment of nanoparticle toxicity profiling. In: Catfish-advances, technology, experiments. IntechOpen. https://doi.org/10.5772/intechopen.107213

Atamanalp M, Kırıcı M, Koktürk M, Kırıcı M, Kocaman EM, Ucar A, Alak G (2023) Polyethylene exposure in rainbow trout; suppresses growth and may act as a promoting agent in tissue-based oxidative response, DNA damage and apoptosis. Process Saf Environ Protect. https://doi.org/10.1016/j.psep.2023.05.005

Bacchetta C, Ale A, Rossi AS, Karakachoff M, Cazenave J (2020) Effects of cold stress on juvenile *Piaractus mesopotamicus* and the mitigation by β-carotene. J Therm Biol 88:102497. https://doi.org/10.1016/j.jtherbio.2019.102497

Baldissera MD, Souza CF, Baldisserotto B, Zimmer F, Paiano D, Petrolli TG, Da Silva AS (2019) Vegetable choline improves growth performance, energetic metabolism, and antioxidant capacity of fingerling Nile tilapia (*Oreochromis niloticus*). Aquaculture 501:224–229. https://doi.org/10.1016/j.aquaculture.2018.11.021

Bard B, Kieffer JD (2019) The effects of repeat acute thermal stress on the critical thermal maximum (CTmax) and physiology of juvenile shortnose sturgeon (*Acipenser brevirostrum*). Can J Zool 97(6):567–572. https://doi.org/10.1139/cjz-2018-0157

Barton BA (2002) Stress in fishes: a diversity of responses with particular reference to changes in circulating corticosteroids. Integr Comp Biol 42(3):517–525. https://doi.org/10.1093/icb/42.3.517

Bej S, Ghosh K, Chatterjee A, Saha NC (2021) Assessment of biochemical, hematological and behavioral biomarkers of *Cyprinus carpio* on exposure to a type-II pyrethroid insecticide Alpha-cypermethrin. Environ Toxicol Pharmacol 87:103717. https://doi.org/10.1016/j.etap.2021.103717

Bojarski B, Witeska M (2020) Blood biomarkers of herbicide, insecticide, and fungicide toxicity to fish—a review. Environ Sci Pollut Res 27:19236–19250. https://doi.org/10.1007/s11356-020-08248-8

Brix KV, De Boeck G, Baken S, Fort DJ (2022) Adverse outcome pathways for chronic copper toxicity to fish and amphibians. Environ Toxicol Chem. https://doi.org/10.1002/etc.5483

Burgos-Aceves MA, Lionetti L, Faggio C (2019) Multidisciplinary haematology as prognostic device in environmental and xenobiotic stress-induced response in fish. Sci Total Environ 670:1170–1183. https://doi.org/10.1016/j.scitotenv.2019.03.275

Corredor-Santamaría W, Serrano Gómez M, Velasco-Santamaría YM (2016) Using genotoxic and haematological biomarkers as an evidence of environmental contamination in the Ocoa River native fish, Villavicencio—Meta, Colombia. Springerplus 5(1):1–10. https://doi.org/10.1186/s40064-016-1753-0

Chang CH, Liu CJ, Lu WJ, Wu LY, Lai KJ, Lin YT, Lee TH (2022) Hypothermal effects on energy supply for ionocytes in gills of freshwater-and seawater-acclimated milkfish, Chanos Chanos. Front Marine Sci. https://doi.org/10.3389/fmars.2022.880103

De Moraes CM, Wanjiku C, Stanczyk NM, Pulido H, Sims JW, Betz HS, Mescher MC (2018) Volatile biomarkers of symptomatic and asymptomatic malaria infection in humans. Proc Natl Acad Sci 115(22):5780–5785. https://doi.org/10.1073/pnas.1801512115

Dettleff P, Zuloaga R, Fuentes M, Gonzalez P, Aedo J, Estrada JM, Valdés JA (2022a) High-temperature stress effect on the red Cusk-Eel (*Geypterus chilensis*) liver: transcriptional modulation and oxidative stress damage. Biology 11(7):990. https://doi.org/10.3390/biology11070990

Dettleff P, Rodríguez J, Patiño-García D, Orellana R, Castro R, Escobar-Aguirre S, Palomino J (2022b) Molecular characterization of embryos with different buoyancy levels in the yellowtail kingfish (*Seriola lalandi*). Animals 12(6):720. https://doi.org/10.3390/ani12060720

El-Agri AM, Emam MA, Gaber HS, Hassan EA, Hamdy SM (2022) Integrated use of biomarkers to assess the impact of heavy metal pollution on *Solea aegyptiaca* fish in Lake Qarun. Environ Sci Eur 34(1):74. https://doi.org/10.1186/s12302-022-00656-6

Farag MR, Attia YA, Al Solami LS, Bovera F, Nizza A, Alagawany M (2022) Behavioral, physiological, and inflammatory responses of Oreochromis niloticus fish exposed to thallium and/or supplementation with *Astragalus membranaceus* polysaccharides. Aquaculture 553:738100. https://doi.org/10.1016/j.aquaculture.2022.738100

Fazio F (2019) Fish hematology analysis as an important tool of aquaculture: a review. Aquaculture 500:237–242. https://doi.org/10.1016/j.aquaculture.2018.10.030

Gutteridge JM, Halliwell B (2018) Mini-Review: Oxidative stress, redox stress or redox success? Biochem Biophys Res Commun 502(2):183–186. https://doi.org/10.1016/j.bbrc.2018.05.045

Ighodaro OM, Akinloye OA (2018) First line defence antioxidants-superoxide dismutase (SOD), catalase (CAT) and glutathione peroxidase (GPX): their fundamental role in the entire antioxidant defence grid. Alexandria J Med 54(4):287–293. https://doi.org/10.1016/j.ajme.2017.09.001

Jeong J, Choi J (2018) Use of adverse outcome pathways in chemical toxicity testing: potential advantages and limitations. Environ Health Toxicol 33(1). https://doi.org/10.5620/eht.e2018002

Khairy EY, Attia MM (2021) Protective effects of vitamin D on neurophysiologic alterations in brain aging: role of brain-derived neurotrophic factor (BDNF). Nutr Neurosci 24(8):650–659. https://doi.org/10.1080/1028415X.2019.1665854

Kokturk M, Yıldırım S, Atamanalp M, Calimli MH, Nas MS, Bolat I, Alak G (2022a) Assessment of oxidative DNA damage, apoptosis and histopathological alterations on zebrafish exposed with green silver nanoparticle. Chem Ecol 38(7):655–670. https://doi.org/10.1080/02757540.2022.2108808

Kokturk M, Yıldırım S, Nas MS, Ozhan G, Atamanalp M, Bolat I, Alak G (2022b) Investigation of the oxidative stress response of a green synthesis nanoparticle (RP-Ag/ACNPs) in zebrafish. Biol Trace Elem Res 200(6):2897–2907. https://doi.org/10.1007/s12011-021-02855-3

Kokturk M, Yıldırım S, Calimli MH, Nas MS, Ibaokurgil F, Ozhan G, Alak G (2023) Perspective on green synthesis of RP-Pd/AC NPs: characterization, embryonic and neuronal toxicity assessment. Int J Environ Sci Technol 20(1):871–882. https://doi.org/10.1007/s13762-022-04005-1

Köktürk M, Yildirim S, Yiğit A, Ozhan G, Bolat İ, Alma MH, Atamanalp M (2022) What is the eco-toxicological level and effects of graphene oxide-boramidic acid (GO-ED-BA NP)?: In vivo study on Zebrafish embryo/larvae. J Environ Chem Eng 10(5):108443. https://doi.org/10.1016/j.jece.2022.108443

Koner D, Banerjee B, Kumari A, Lanong AS, Snaitang R, Saha N (2021) Molecular characterization of superoxide dismutase and catalase genes, and the induction of antioxidant genes under the zinc oxide nanoparticle-induced oxidative stress in air-breathing magur catfish (*Clarias magur*). Fish Physiol Biochem 47:1909–1932. https://doi.org/10.1007/s10695-021-01019-3

Kumar M, Sabu S, Sangela V, Meena M, Rajput VD, Minkina T, Vinayak V (2023) The mechanism of nanoparticle toxicity to cyanobacteria. Arch Microbiol 205(1):1–10. https://doi.org/10.1007/s00203-022-03370-2

Lesniak A, Poznański P, Religa P, Nawrocka A, Bujalska-Zadrozny M, Sacharczuk M (2021) Loss of brain-derived neurotrophic factor (BDNF) resulting from congenital-or mild traumatic brain injury-induced blood–brain barrier disruption correlates with depressive-like behaviour. Neuroscience 458:1–10. https://doi.org/10.1016/j.neuroscience.2021.01.013

Li L, Liu Z, Quan J, Lu J, Zhao G, Sun J (2022) Metabonomics analysis reveals the protective effect of nano-selenium against heat stress of rainbow trout (Oncorhynchus mykiss). J Proteomics 259:104545. https://doi.org/10.1016/j.jprot.2022.104545

Manti S, Xerra F, Spoto G, Butera A, Gitto E, Di Rosa G, Nicotera AG (2023) Neurotrophins: expression of brain-lung axis development. Int J Mol Sci 24(8):7089. https://doi.org/10.3390/ijms24087089

Mehrotra P (2016) Biosensors and their applications–A review. J Oral Biol Craniofac Res 6(2):153–159

Murer MG, Yan Q, Raisman-Vozari R (2001) Brain-derived neurotrophic factor in the control human brain, and in Alzheimer's disease and Parkinson's disease. Prog Neurobiol 63(1):71–124. https://doi.org/10.1016/S0301-0082(00)00014-9

Nan S, Wang Y, Xu C, Wang H (2021) Interfering microRNA-410 attenuates atherosclerosis via the HDAC1/KLF5/IKBα/NF-κB axis. Molecular Therapy-Nucleic Acids 24:646–657. https://doi.org/10.1016/j.omtn.2021.03.009

Ogamba EN, Iyiola AO, Yarkwan B, Adetola BO (2023) Potentials, threats, and sustainable conservation strategies of plankton and macrophytes. In: Sustainable utilization and conservation of Africa's biological resources and environment. Springer Nature Singapore, Singapore, pp 85–117. https://doi.org/10.1007/978-981-19-6974-4_4

Olivares-Rubio HF, Espinosa-Aguirre JJ (2021) Acetylcholinesterase activity in fish species exposed to crude oil hydrocarbons: a review and new perspectives. Chemosphere 264:128401. https://doi.org/10.1016/j.chemosphere.2020.128401

Özgeriş FB, Ycltckin AÇ, Ucar A, Çağlar Ö, Parlak V, Arslan ME, Alak G (2023) Toxic releases and exposure assessment: a multi-endpoint approach in fish for ferrocene toxicity. Process Saf Environ Prot 169:636–645. https://doi.org/10.1016/j.psep.2022.11.052

Parlak V, Ozgeris B, Ucar A, Cilingir Yeltekin A, Ozgeris FB, Cağlar O, Atamanalp M (2022) Hematotoxic, oxidative and genotoxic damage in rainbow trout (*Oncorhynchus mykiss*) after exposure to 3-benzoylpyridine. Toxicol Mech Methods 32(7):501–509. https://doi.org/10.1080/15376516.2022.2049413

Patel DM, Brinchmann MF, Hanssen A, Iversen MH (2022) Changes in the skin proteome and signs of allostatic overload type 2, chronic stress, in response to repeated overcrowding of Lumpfish (*Cyclopterus lumpus* L.). Front Mar Sci 9. https://doi.org/10.3389/fmars.2022.891451

Petitjean Q, Jean S, Gandar A, Côte J, Laffaille P, Jacquin L (2019) Stress responses in fish: from molecular to evolutionary processes. Sci Total Environ 684:371–380. https://doi.org/10.1016/j.scitotenv.2019.05.357

Pozuelo-Campos S, Casero-Alonso V, Amo-Salas M (2023) Optimal designs for detecting and characterizing hormesis in toxicological tests. Chemom Intell Lab Syst 235:104753. https://doi.org/10.1016/j.chemolab.2023.104753

Sakai M, Hikima JI, Kono T (2021) Fish cytokines: current research and applications. Fish Sci 87:1–9. https://doi.org/10.1007/s12562-020-01476-4

Schreck CB, Tort L (2016) The concept of stress in fish. In Fish physiology. Academic Press, vol 35, pp 1–34. https://doi.org/10.1016/B978-0-12-802728-8.00001-1

Serafim A, Company R, Lopes B, Fonseca VF, França S, Vasconcelos RP, Cabral HN (2012) Application of an integrated biomarker response index (IBR) to assess temporal variation of environmental quality in two Portuguese aquatic systems. Ecol Ind 19:215–225. https://doi.org/10.1016/j.ecolind.2011.08.009

Serafini S, de Freitas SC, Baldissera MD, Baldisserotto B, Segat JC, Baretta D, da Silva AS (2019) Fish exposed to water contaminated with eprinomectin show inhibition of the activities of AChE and Na+/K+-ATPase in the brain, and changes in natural behavior. Chemosphere 223:124–130. https://doi.org/10.1016/j.chemosphere.2019.02.026

Spagnuolo MS, Bergamo P, Crescenzo R, Iannotta L, Treppiccione L, Iossa S, Cigliano L (2020) Brain Nrf2 pathway, autophagy, and synaptic function proteins are modulated by a short-term fructose feeding in young and adult rats. Nutr Neurosci 23(4):309–320. https://doi.org/10.1080/1028415X.2018.1501532

Topal A, Özdemir S, Arslan H, Çomaklı S (2021) How does elevated water temperature affect fish brain? (A neurophysiological and experimental study: assessment of brain derived neurotrophic factor, cFOS, apoptotic genes, heat shock genes, ER-stress genes and oxidative stress genes). Fish Shellfish Immunol 115:198–204. https://doi.org/10.1016/j.fsi.2021.05.002

Ucar A, Yeltekin AÇ, Köktürk M, Calimli MH, Nas MS, Parlak V, Atamanalp M (2023) Has PdCu@ GO effect on oxidant/antioxidant balance? Using zebrafish embryos and larvae as a model. Chem Biol Interact 378:110484. https://doi.org/10.1016/j.cbi.2023.110484

Ucar A, Parlak V, Ozgeris FB, Yeltekin AC, Arslan ME, Alak G, Atamanalp M (2022) Magnetic nanoparticles-induced neurotoxicity and oxidative stress in brain of rainbow trout: mitigation by ulexite through modulation of antioxidant, anti-inflammatory, and antiapoptotic activities. Sci Total Environ 838:155718. https://doi.org/10.1016/j.scitotenv.2022.155718

Uçar A, Parlak V, Çilingir Yeltekin A, Özgeriş FB, Çağlar Ö, Türkez H, Atamanalp M (2021) Assesment of hematotoxic, oxidative and genotoxic damage potentials of fipronil in rainbow trout Oncorhynchus mykiss, Walbaum. Toxicol Mechan Methods 31(1):73–80. https://doi.org/10.1080/15376516.2020.1831122

Umar AM, Aisami A (2020) Acetylcholinesterase enzyme (AChE) as a biosensor and biomarker for pesticides: a mini review. Bull Environ Sci Sustain Manage 4(1):7–12. e-ISSN 2716-5353. https://doi.org/10.54987/bessm.v4i1.526

Yeltekin AÇ, Ucar A, Parlak V, Özgeriş FB, Türkez H, Esenbuğa N, Alak G (2022) Borax exerts protective effect against ferrocene-induced neurotoxicity in Oncorhynchus mykiss. J Trace Elem Med Biol 72:126996. https://doi.org/10.1016/j.jtemb.2022.126996

Yuan J, Wang D, Liu Y, Chen X, Zhang H, Shen F, Fu J (2018) Hydrogen-rich water attenuates oxidative stress in rats with traumatic brain injury via Nrf2 pathway. J Surg Res 228:238–246. https://doi.org/10.1016/j.jss.2018.03.024

Zhang T, Yang M, Pan H, Li S, Ren B, Ren Z, Ma J (2017) Does time difference of the acetylcholinesterase (AChE) inhibition in different tissues exist? A case study of zebra fish (Danio rerio) exposed to cadmium chloride and deltamethrin. Chemosphere 168:908–916. https://doi.org/10.1016/j.chemosphere.2016.10.119

Wang Y, An R, Umanah GK, Park H, Nambiar K, Eacker SM, Dawson TM (2016) A nuclease that mediates cell death induced by DNA damage and poly (ADP-ribose) polymerase-1. Science 354(6308):aad6872. https://doi.org/10.1126/science.aad6872

Chapter 9
Nanotoxicology

Gonca Alak

Abstract Today, the field of nanotechnology is developing incredibly. With this rise of nanotechnology, the potential risks of nanomaterials/nanoparticles to aquatic organisms have led to growing concerns, along with environmental issues. Due to their anthropogenic release, these nanoparticles (NPs) are seen as a potential hazard to aquatic life and aquatic environmental health. In recent years, the interaction between NPs and biological processes has received increasing attention, and the need for research into their presence, characterization and toxicological effects on ecosystem components/environments, is increasing. This section will contribute to the sustainable improvement of existing data, along with recommendations on how to support the growth of a safe and sustainable nanotechnology industry, what new potential threats could be posed by the widespread use of nanoparticles, raise awareness of the discipline of aquatic nanotoxicology, and how to prevent this threat.

Keywords Nanotoxicology · Fish · Green synthesis · Metal nanoparticles

List of Abbreviations

Nm	Nanometer
NP/NPs	Nanoparticles
NMs	Nanomaterials
MNs	Metallic nanomaterials
ROS	Reactive oxygen groups
DNA	Deoxyribonucleic acid
8-OHdG	8-Hydroxy-2′-deoxyguanosine

G. Alak (✉)
Department of Seafood Processing Technology, Faculty of Fisheries, Atatürk University, Erzurum, Türkiye
e-mail: galak@atauni.edu.tr

M. Atamanalp et al., *Aquatic Toxicology in Freshwater*, Springer Water,
https://doi.org/10.1007/978-3-031-56669-1_9

9.1 Introduction

Nanotechnology; is defined as a multidisciplinary technology that combines many research areas or disciplines, allows research and applications at the atomic, molecular and macromolecular levels. It is the art of processing and controlling matter at the nanoscale (Shah et al. 2021). Research at the nanoscale is called nanoscience. Nano is a unit of length measurement that refers to a billionth of a meter or a millionth of a millimeter. A nanometer (nm) is the length obtained by lining up 2–3 atoms side by side. Work at the nanoscale is concerned with developing materials and devices with a size of 1–100 nm. The application of nanotechnology allows changes to be made in the basic physical and chemical properties of conventional materials as their size changes to the nanoscale. The properties and behaviors that a substance has at the macro scale are different from those at the nano scale. Nanoparticles have unique optical, magnetic and chemical properties, especially at sizes smaller than 20 nm. While much is known about the composition and position of particulate matter, it is difficult to learn about its detailed size distribution. This category is interesting for nanotoxicology field, as NPs are defined as particles with at least two dimensions below 100 nm (i.e. 0.1 mm) (Alak et al. 2023a, b).

NPs are produced during abiotic and biotic chemical processes (combustion processes and various natural forms). Since the usages in many applications in aerospace, space exploration, defense, machinery, construction industries, electronics, materials industry, computer technologies, automobile, metal industry, environment, energy, food, agriculture, chemical engineering, water treatment, textile, cosmetics, medicine and health, their commercial production has increased steadily as materials are likely to be developed. Nanomaterials (NMs) with various properties have begun to be synthesized for use in these areas. At the head of this century, there was a growing interest in the emerging fields of nanoscience and nanotechnology (Hulla et al. 2015; Alak et al. 2023a), and the cost of using nanotechnology for more than 1300 products had reached 50 billion dollars (Dar et al. 2020). In the European NMs market, it had been reported that more than $ 2.5 million in revenue was generated in 2015, and it is expected to reach approximately $ 9 million in revenue by 2022 (Primožič et al. 2021). It seems that in the future, the load of these particles will increase in ecosystems (atmosphere, water and soil) and therefore in their components (human, plant and animal).

9.2 Bibliographic Extraction

The following keyword search string was entered into the search bar on the Scopus website (https://www.scopus.com/) [TITLE-ABS-KEY ("nanotoxicology") AND [TITLE-ABS-KEY ("aquatic ecosystem, fish, aquatic toxicology")]. Regardless of the years, languages, sources, or documents, this bibliographic investigation produced 490 research documents [Article (405), Review (59), Book Chapter (12), Conference Paper (7), Book (3), Editorial (3), Short Survey (2)] from 2014 to 2023

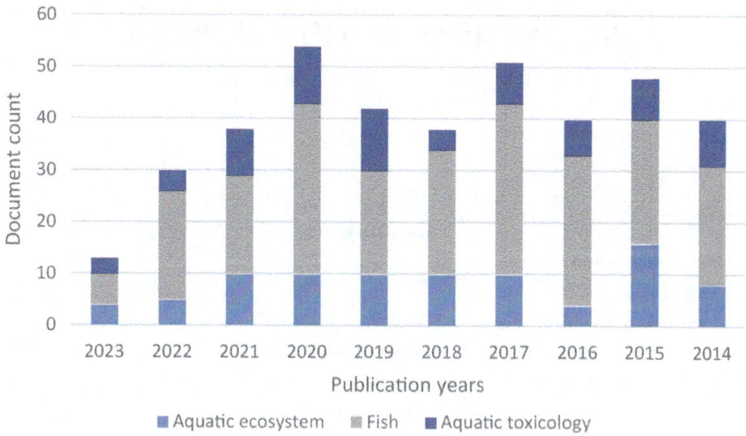

Fig. 9.1 Studies by year and area

were all included in this collection of materials. The number of published documents and years showed in Fig. 9.1. The fact that there are so many research papers in this area is due to the growing interest in both the academic and professional societies. The distribution of publications in the field of nanotechnology is given in Fig. 9.1 on the basis of keywords. In this ranking, when sorting from the most publications to the least, the ranking of fish (260), aquatic ecosystem (81), and aquatic toxicology (64) draws attention.

9.3 Synthesis of Nanoparticles

The synthesis of nanoparticles by physical and chemical methods is called traditional methods because they have been known and applied for many years. The ways to synthesize a nanomaterial are shown in Fig. 9.2 (Sahoo et al. 2021). Traditional synthesis methods are examined in two basic approaches: top-down and bottom-up. In the top-down synthesis method, macro-sized materials are ground by mechanical milling and nano-sized materials can be obtained. In the bottom-up approach, chemical methods such as gas condensation, sol–gel technique, vacuum application, hydrolysis are used (Mutaf et al. 2023). In these production methods, physical, chemical and biological methods are applied for different sizes and shapes. Low-cost production of nanoparticles on large scales is possible using physical and chemical methods, so they are often preferred in scientific research and industrial applications.

Metallic nanomaterials (MNs), classified under nanomaterials, are widely used in many fields due to their exceptional physical and chemical properties. With their introduction of a wide range of innovative solutions, they are attracting more and more attention day by day and commercial products developed in the light of scientific studies have been increasing in recent years (Mutaf et al. 2023). The increasing

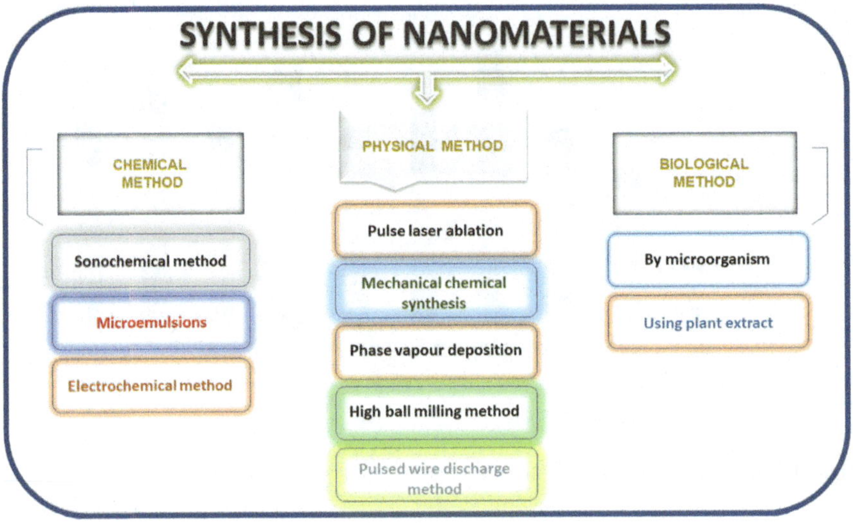

Fig. 9.2 Nanomaterial synthesis methods

MNs application in industries and society has led to a rapid growth in the production volume of NMs, 80% of which are metals or metal oxides, with an annual production of approximately 242,000 tons (Zhou et al. 2023). Metal oxide nanoparticles have wide applications in the field of technology with their semiconductor, electroluminescent or thermoelectric properties. They are also used as drug delivery systems for treatment and diagnostics in biomedical and in environmental decontamination applications. Classical methods of obtaining metal oxide nanoparticles are based on chemical and physical techniques using high energy input and hazardous and expensive chemicals that have a negative impact on the environment. The production of metal oxide nanoparticles through biogenic synthesis has recently received increasing attention as it is a new process for the development of engineering materials (Alak et al. 2023b).

The biosynthesis of NPs is more advantageous than chemical and physical methods due to its environmental friendliness, commercial viability and economy (Alak et al. 2023b). Plant extracts are of great interest in the green synthesis of nanoparticles. Plants provide positive manipulation and control over crystal growth and stabilization. Therefore, plant-mediated synthesis of silver nanoparticles is highly effective in achieving the desired shape, size and difference. Plants are widely used for the synthesis of metal composition nanoparticles due to their easy availability, low cost, content of important bioactive compounds, non-toxic and environmental friendliness. The synthesis of metal composition nanoparticles using aromatic and medicinal plants has revealed many wide range of biological activities (anticancer, antioxidant, antibacterial, antiviral, antifungal, anti-inflammatory activities) (Gecer et al. 2021).

The plant-based nanoparticle synthesis procedure is easier and a metal salt is synthesized with the targeted plant extract. Green nanoparticle synthesis is completed in a short time interval (minutes to several hours) at room temperature (Gour and Jain 2019). This strategy has received much more attention in recent years, especially for silver (Ag) and gold (Au) NPs, which are safer compared to other metallic NPs (Alkhalaf et al. 2020; Tailor et al. 2020; Nayem et al. 2020; Hosseinzadeh et al. 2020).

Thanks to the green method, NPs with different shapes, sizes, contents and physicochemical properties can be synthesized. The green synthesis of NPs involves the use of prokaryotic/eukaryotic cells or extracted biomolecules that act as reducing agents (Jeevanandam et al. 2016; Köktürk et al. 2023). Plant biomass/extract, from a variety of biological materials, has several inert advantages over other microscopic organisms in the synthesis of NP. Plant-mediated biosynthesis of metallic NPs occurs through biomolecules contained in plant biomass that contain organic functional groups (Köktürk et al. 2022a, 2023a; b, c). Various phytochemicals obtained from the leaves, roots and other stems of plants play a reducing role by converting into quinone form in the presence of reactive oxygen groups (ROS) (Thakur and Karak 2012; Thakur et al. 2020). It plays an important role in the synthesis of NPs and the desired characteristic properties in the technical and environmental factors used (such as temperature, pH, concentration of reducing agents, metal ion concentration, interaction time, pressure) as well as the synthesis method (Rana et al. 2020; Çiftçi et al. 2021).

Although there are various studies on the synthesis of nanoparticles by bacteria, the synthesis mechanism has not been fully elucidated. The synthesis is categorized as intracellular and extracellular according to its location (Yavuz and Yılmaz, 2021). Recently, there are approaches to using plants as a new alternative due to the limitations associated with the use of bacteria in synthesizing nanoparticles (Mehwish et al. 2021). In this approach, plants are recognized as one of the richest sources for the green synthesis of nanomaterials (Zhao et al. 2021). The use of plant materials such as leaves, stems, fruits, bark, roots and flowers in green synthesis makes an important contribution to both non-toxic, environmentally friendly and capping or stabilizing agents (Aravind et al. 2021a, b; Gul et al. 2021). In the green synthesis of silver nanoparticles of different sizes and shapes, various plants such as *Laminaria japonica, Azadirachta indica, Emblica officinalis, Tribulus terrestris, Pimpinella anisum* L., *Chenopodium album, Taxus baccata, Fritillaria çiçeği, Trifolium resupinatum, Solanum melongena, Cardiospermum halicacabum L., Olea europaea, Ziziphora tenuior L, Justicia adhatoda, Nasturtium officinale, Syzygium cumini* are used (Damirchi et al. 2020; Iliger et al. 2021; Mehwish et al. 2021). The use areas of biological nanoparticles are quite wide and continue to evolve. In Fig. 9.3, a design is made according to different application areas.

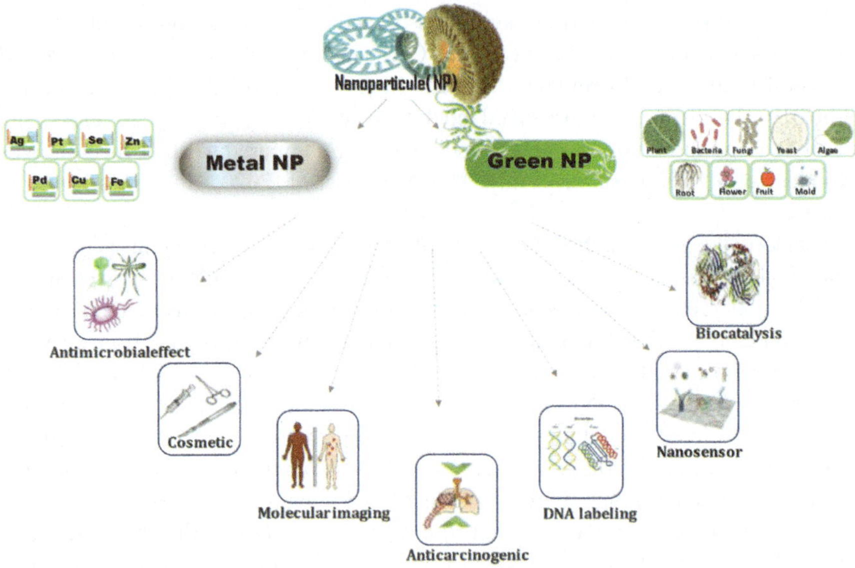

Fig. 9.3 Uses of nanoparticles

9.4 Physical/Chemical Properties of Nanoparticles

NPs are broadly divided into several categories based on their morphology, size, and chemical properties. Some well-known NP classes based on their physical and chemical properties are given in Fig. 9.4 (Khan et al. 2019). The obvious properties of nanoparticles that need to be evaluated before any in vivo or in vitro testing are properties such as shape, chemical composition, specific surface area, primary particle size, size distribution of particles, aggregation state, and elemental composition, mostly encompassing morphological and metric-related properties. While all of these features are relevant to any toxicity study, the choice of features to be more accurately measured should be tailored to the endpoint investigated in that study (Fubini et al. 2010). In this direction, the following features are evaluated.

Aggregation status; NPs that tend to aggregate may retain their functionality, but an increase in their size may result in decreased uptake by cells. Elemental composition; different particle composition leads to different behavior effects. Mass concentration; normally an increase in contaminant concentration causes an increase in toxicity/ effect, which is not always the case for NPs. Particle number concentration; NPs have low mass concentrations, but total particle numbers show a high percentage. NP Shape; different particle shapes have different affinities or accessibility, for example, the transport of NPs from membranes to cells and the different antibacterial behavior of these NPs. Size and size distribution; NPs are defined and classified according to their size because NP size is one of the key characteristics that defines moving behaviors. Resolution; Ionic forms of soluble NPs can be harmful or toxic. Speciation; Different strains may show different behavior, toxicity and efficacy. Structure; It can

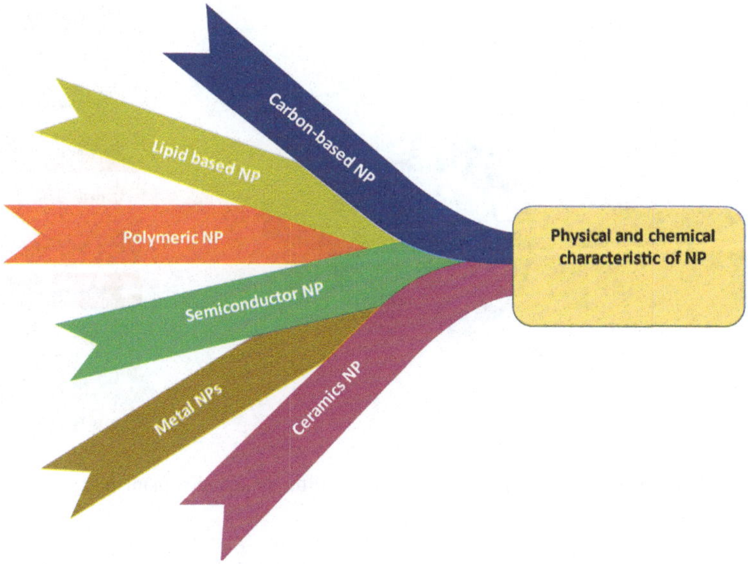

Fig. 9.4 Characterizations of nanoparticles

affect NP stability or behavior. Surface area and porosity; the increase in surface area increases reactivity and absorption behavior. Surface load; surface charge has an effect on NP stability, especially in dispersions. Surface chemistry; coatings can consist of different chemical compositions, and this can affect NP behavior or toxicity. NPs represent an intermediate supramolecular state of matter between bulk and molecular material. Apart from the particle size providing a very large surface-to-volume ratio, NP biocompatibility depends on the surface properties, the charges carried by the NP, and the chemical reaction. Polymeric macromolecules interact strongly with cell membranes in vitro. This interaction of NPs with the surface lining layers of biological tissues is determined by the surface chemistry and reactivity of NPs.

9.5 Toxicity of Nanoparticles

In this section, we sought to assess the need for future research on toxicological effects, with special emphasis on aquatic organisms, including the formation and characterization of NPs, as well as transport processes, environment interactions, and material-specific responses. Considering the fact that aquatic organisms represent by far the largest interface between the environment and the hydrosphere consisting of resources on land, especially oceans and seas, rivers, lakes and groundwater, it has been observed that this important subject is still not adequately explored.

Physical and chemical approaches to synthesizing NPs suggest that nanporticles may cause stress on organisms due to their toxic metabolites (Valerio-García et al.

Fig. 9.5 Ways of nanoparticles' adversely effects on different aquatic organisms

2017; Tai et al. 2019) (Fig. 9.5.). MNs released into the aquatic environment will pose a risk to aquatic organisms such as fish, crustaceans, algae, bacteria (Zhou et al. 2023). Although it is known that nanoparticles can overcome some mechanical barriers created by a biological system due to their similarity to cellular proteins and components (Uçar et al. 2022, 2023), it is still largely unknown how NPs will behave in any environment.

However, some reasonable predictions can be made about their potential hazards based on knowledge of their dimensions, surface charges and chemical reactivity briefly mentioned above. But since they are basically supermolecular entities, their biological interactions cannot be easily predicted, as new unpredictable properties may emerge based on information from their surface properties (Gürmen et al. 2008). Risk assessment of NPs used in various applications is needed to ensure the sustainable development of nanotechnology (Atamanalp et al. 2022). Such risk assessments require appropriate analyses and methods to perform both impact and exposure assessments (Uçar et al. 2022, 2023; Köktürk et al. 2023; Alak et al. 2023a, b). Traditionally, exposure assessment is recommended to include both a modeling and a measurement approach, both approaches requiring instrumentation and analytical techniques (Dağlıoğlu 2018a) Estimation of NP's environmental concentrations of through modeling is based on emission scenarios (e.g. from production volumes and life cycle assessments) and degradation parameters (e.g. fate and behavior). Analytical techniques are sensitive enough to measure low-concentration NPs, as the small particles normally represent only a small fraction of the total mass. In ecotoxicity studies, nanoparticle characterization has become very important since the basic parameters that affect/change the biological activity of NPs are largely unknown (Dağlıoğlu 2018b). Some physico-chemical properties have been found to be related to toxicity, such as sustained release of ROS, radical quenching, surface active metal ions, structural defects. These are clues for understanding the molecular mechanisms of toxicity and ultimately for the safe production of engineered nanomaterials. These clues should not be included in the "characterization" as they can be used to screen for nanotoxicity. The materials under investigation must be modified in such properties during the course of the study to discover the links between chemical properties and the resulting biological reactions.

A. Toxicity pathways of nanoparticules

In traditional methods, reducing and stabilizing chemicals are used for reduction of metal ions, surface modification after nanoparticle formation and preservation of stability. Since the toxic chemicals used are carcinogenic and allergenic, they limit the biotechnological usage areas of nanoparticles after production. In addition, high volumes of chemicals released as waste from commercial productions harm the ecosystem. This situation will bring a great danger to the living population and biodiversity (Mutaf et al. 2023).

Interestingly, it is now recognized by both the scientific community and regulatory agencies that physico-chemical data should be included in any toxicology study conducted. However, recent studies and reviews of the genotoxicity of engineered nanomaterials (Boczkowski and Hoet 2010) highlight the role of new physico-chemical aspects and physico-chemical data that should be considered in nanotoxicology studies.

Several new physico-chemical procedures have been proposed to assess NP toxicity. The use of the antioxidant potential of serum and X-ray fluorescence microscopy are used to determine elemental distribution in cells following NP internalization and to assess oxidative damage after exposure to NPs.

Many other detailed reviews of the NPs genotoxicity (Atamanalp et al. 2022; Köktürk et al. 2022a, b, c, 2023; Alak et al. 2023a; Uçar et al. 2023) clearly show that size is important, but not everything. Therefore, the toxic potential may change by reducing the size of the particles, but the threshold value depends on the chemical nature of the particles. Most genotoxicity studies on nanomaterials have generally used classical genotoxicity testing methods. In most cases, this approach does not take into account that NPs have properties that make them different from any molecular toxicant, but also from the same base material with larger particle size (Köktürk et al. 2022a; b, c). Despite this, several physico-chemical properties have been proven in processes that cause nanoparticle deoxyribonucleic acid (DNA) damage, the most relevant being ROS formation, surface-exposed metal ions, surface charge, and structural defects. In addition, size, shape and degree of hydrophilicity/hydrophobicity play an important role in genotoxicity by regulating particle uptake (Atamanalp et al. 2022; Alak et al. 2023a; Köktürk et al. 2023).

Before the particles interact with the cells, they will adsorb different molecules from the surrounding fluids and be partially or completely covered mainly by proteins (step 1, particle coverage). This process depends on both the particle surface functionalities and properties of the protein (Fubini et al. 2010; Abbasi et al. 2023; Setia et al. 2023). After entering the particle uptake site may damage adjacent cells via a direct or indirect mechanism (step 2, primary genotoxicity) or be engulfed by immune system cells (typically alveolar macrophages), resulting in persistent inflammation, in turn causing other damages cells (step 3, secondary genotoxicity).

Determining the cytotoxic effects of NPs with current methods like other xenobiotics has gained importance in recent years and it has been found that most of the metallic nanoparticles affect protein synthesis, damage DNA and affect the electron transport chain. Oxidative damage to DNA caused by free radicals such as hydroxyl radicals; It causes base damage, chain breaks, and changes that include multiple disorders such as DNA protein cross-linking. Although many different oxidative damage products caused by agents by different mechanisms have been identified, 8-OHGua with guanine derivative damage and its deoxynucleoside 8-hydroxy-2'-deoxyguanosine (8-OHdG) have been the subject of many studies. In these studies, it has been reported that oxidative stress may cause damage to macromolecules such as lipid, protein and DNA, and genotoxic effect by creating structural damage to DNA (Alak et al. 2018, 2019, 2020; Atamanalp et al. 2021). In addition, it has been reported that cells may go to necrosis as a result of triggering immunological mechanisms and the release of certain cytokines and mediators from the immune system (Erkekoğlu and Baydar 2021).

In toxicity studies, the effects of any chemical on tissues and cells can be studied by various methods. In in vivo studies, which is one of these methods, the responses of organisms to chemicals are realized by the dynamic interaction of tissue and organ systems (Alak et al. 2020; Uçar et al. 2022; Atamanalp et al. 2021). In addition to these toxicology studies, studies for different purposes (especially the development of various diseases and the discovery of metabolic pathways in recent years, drug designs and treatments suitable for them) have accelerated in in vivo trials in which fish are used as model organisms (Köktürk et al. 2020, 2021; Alak et al. 2020, 2023a).

2 Novelty perspectives of nanoparticules

While studies in the field of nanotechnology are increasing rapidly, there are some concerns about the possible risks and effects of the created nanomaterials on the environment and human health (Amooaghaie et al. 2015). These concerns have led to the development of environmentally friendly green synthesis methods. NPs produced from biological materials are known as biogenic NPs, and the associated synthesis process is called green synthesis. Emphasizing the importance of meeting nanotechnology and biotechnology, biological synthesis of inorganic NP has given great importance to environmentally friendly technologies in nanomaterial production (Azizi et al. 2017, 2020). It has been realized that with the green synthesis method, methodologies that reduce or neglect the use or production of raw materials, products, by-products, solvents, reagents and the like, which can significantly impair human health and the environment, can be used (Atamanalp et al. 2022; Köktürk et al. 2022a, b, c, 2023; Alak et al. 2023a; Uçar et al. 2023). For this reason, green-medium methods have been used in recent years for the preparation of small-sized and crystalline metal oxide nanoparticles. Advantages such as efficient particle production, high success, low cost and ease of characterization (Köktürk et al. 2022a; b, c) have significantly enhanced green synthesis methods due to their potential to reduce the toxicity of NPs.

Nanotoxicology emerges as one of the bionanosystems sciences dealing with the study and application of the toxicity of nanomaterials. Nanotoxicological studies are used to determine to how and to what extent these properties pose a threat to the environment and human health. Many studies have shown that some nanoparticles have a toxic effect on biological systems. But scientific knowledge on the potentially harmful effects of nanoparticles lags behind nanotechnology development (Atamanalp et al. 2022; Uçar et al. 2022; Köktürk et al. 2022a, b, c, 2023; Alak et al. 2023a, b; Uçar et al. 2023).

The more detailed characteristics of the parameters/evidence regarding the route of exposure and how organisms cope with this internalization and subsequent toxicity cannot be explained due to the lack of data and scientific knowledge on the subject (Zhou et al. 2023).

9.6 Conclusion

From the studies examined, it can be concluded that nanoparticles affect the growth rate, biochemical macromolecule and general physiology of aquatic organisms and even lead to apoptosis. However, new modelling is needed for the behavior of NMs in the organism's bodies, biodistribution dynamics, and impact targets. Strict regulatory guidelines for the use and release of nanoparticles must be established and enforced by government agencies to protect aquatic environments on a global scale. The main mechanism of nanoparticle toxicity, (1) morphological changes in aquatic organisms (2) tissue and system-based damages (3) Oxidative stress leading to the formation of ROS, and changes in antioxidant enzyme, cytokines, immunological parameters, DNA damage and apoptosis processes in response to this stress.

In the light of the information summarized above, it seems that exposure time and hydrodynamic diameter of NPs play a critical role in toxicity activity. In order to reduce and control these effects, it seems important to make legal arrangements, monitor water parameters, determine the entry patterns of nanoparticles in aquatic environments, and the concentration limits in aquatic ecosystems, as well as taking into account the parameters related to the exposure route. However, there is an urgent need for the development of new descriptors for multi-species modeling studies that can accurately reveal the difference between species in terms of toxicity mechanism.

Combining and systematizing existing morphological and cytological research methods and establishing new techniques for propagating organoids composed of several cell types will fully unlock the potential of existing models for nanotoxicological studies and improve their quality. Because still, many organs and tissues still lack adequate organoid analogues.

Acknowledgements Many thanks to my colleagues from Fisheries Faculty who supported me in the research for their help Atatürk University for their support in my academic studies.

References

Abbasi R, Shineh G, Mobaraki M, Doughty S, Tayebi L (2023) Structural parameters of nanoparticles affecting their toxicity for biomedical applications: a review. J Nanopart Res 25(3):43. https://doi.org/10.1007/s11051-023-05690-w

Alak G, Parlak V, Ucar A, Yeltekin AC, Ozgeris FB, Caglar O, Turkez H (2020) Oxidative and DNA damage potential of colemanite on zebrafish: brain, liver and blood. Turk J Fish Aquat Sci 20(8):593–602. https://doi.org/10.4194/1303-2712-v20_8_02

Alak G, Ucar A, Yeltekin AÇ, Çomaklı S, Parlak V, Taş IH, ... Türkez H (2018) Neuroprotective effects of dietary borax in the brain tissue of rainbow trout (Oncorhynchus mykiss) exposed to copper-induced toxicity. Fish Physiol Biochem 44:1409–1420

Alak G, Parlak V, Yeltekin AÇ, Ucar A, Çomaklı S, Topal A, ... Türkez H (2019) The protective effect exerted by dietary borax on toxicity metabolism in rainbow trout (Oncorhynchus mykiss) tissues. Comp Biochem Physiol C Toxicol Pharmacol 216:82–92

Alak G, Yeltekin AÇ, Köktürk M, Nas MS, Parlak V, Calimli MH, Atamanalp M (2023a) May PdCu@ f-MWCNT NPs be an ecotoxicologic risk? Appl Organomet Chem 37(3):e7013. https://doi.org/10.1002/aoc.7013

Alak G, Atamanalp M, Parlak V, Ucar A (2023b) Ecent insight into nanotechnology in fish processing: a knowledge gap analysis. Nanomaterial. https://doi.org/10.1088/1361-6528/acda9e

Alkhalaf MI, Hussein RH, Hamza A (2020) Green synthesis of silver nanoparticles by Nigella sativa extract alleviates diabetic neuropathy through anti-inflammatory and antioxidant effects. Saudi J Biol Sci 27(9):2410–2419. https://doi.org/10.1016/j.sjbs.2020.05.005

Amooaghaie R, Saeri MR, Azizi M (2015) Synthesis, characterization and biocompatibility of silver nanoparticles synthesized from Nigella sativa leaf extract in comparison with chemical silver nanoparticles. Ecotoxicol Environ Saf 120:400–408. https://doi.org/10.1016/j.ecoenv.2015.06.025

Aravind M, Ahmad A, Ahmad I, Amalanathan M, Naseem K, Mary SMM, Zuber M (2021a) Critical green routing synthesis of silver NPs using jasmine flower extract for biological activities and photocatalytical degradation of methylene blue. J Environ Chem Eng 9(1):104877. https://doi.org/10.1016/j.jece.2020.104877

Aravind M, Amalanathan M, Mary MSM (2021b) Synthesis of TiO_2 nanoparticles by chemical and green synthesis methods and their multifaceted properties. SN Appl Sci 3:1–10. https://doi.org/10.1007/s42452-021-04281-5

Atamanalp M, Parlak V, Özgeriş FB, Çilingir Yeltekin A, Ucar A, Keleş MS, Alak G (2021) Treatment of oxidative stress, apoptosis, and DNA injury with N-acetylcysteine at simulative pesticide toxicity in fish. Toxicol Mech Methods 31(3):224–234. https://doi.org/10.1080/15376516.2021.1871794

Atamanalp M, Ucar A, Alak G (2022) Catfish as an ecotoxicological model for assessment of nanoparticle toxicity profiling. In: Catfish-advances, technology, experiments. IntechOpen. https://doi.org/10.5772/intechopen.107213

Azizi N, Mohammadi T, Behbahani RM (2017) Comparison of permeability performance of PEBAX-1074/TiO2, PEBAX-1074/SiO2 and PEBAX-1074/Al2O3 nanocomposite membranes for CO2/CH4 separation. Chem Eng Res Des 117:177–189. https://doi.org/10.1016/j.cherd.2016.10.018

Azizi-Lalabadi M, Alizadeh-Sani M, Divband B, Ehsani A, McClements DJ (2020) Nanocomposite films consisting of functional nanoparticles (TiO_2 and ZnO) embedded in 4A-Zeolite and mixed polymer matrices (gelatin and polyvinyl alcohol). Food Res Int 137:109716. https://doi.org/10.1016/j.foodres.2020.109716

Boczkowski J, Hoet P (2010) What's new in nanotoxicology? Implications for public health from a brief review of the 2008 literature. Nanotoxicology 4(1):1–14. https://doi.org/10.3109/17435390903428844

Çiftçi E (2021) Distilled water-based AlN+ ZnO binary hybrid nanofluid utilization in a heat pipe and investigation of its effects on performance. Int J Thermophys 42(3):38. https://doi.org/10.1007/s10765-021-02792-2

Dağlıoğlu C (2018a) İlaç taşıma sistemleri olarak nanopartiküller kullanılarak pasif ve aktif tümör hedeflemelerinin karşılaştırmalı incelenmesi. Academic Platform-J Eng Sci

Dağlıoglu Y (2018) Nanopartikül karakterizasyon yöntemleri ve ekotoksisite deneylerindeki önemi. Marmara Fen Bilimleri Dergisi 30(1):1–17. https://doi.org/10.7240/marufbd.346547

Damirchi TZ, Rostami Charati F, Akbari R, Daneshvar A (2020) Green synthesis of silver nanoparticles using the aqueous extract of Viscum album fruit. Nanochem Res 5(1):104–110. https://doi.org/10.22036/NCR.2020.01.010

Dar AH, Rashid N, Majid I, Hussain S, Dar MA (2020) Nanotechnology interventions in aquaculture and seafood preservation. Crit Rev Food Sci Nutr 60(11):1912–1921. https://doi.org/10.1080/10408398.2019.1617232

Erkekoğlu P, Baydar T (2021) Güncel in vitro sitotoksisite testleri. Hacettepe Univ J Facul Pharm 41(1):45–63

Fubini B, Ghiazza M, Fenoglio I (2010) Physico-chemical features of engineered nanoparticles relevant to their toxicity. Nanotoxicology 4(4):347–363. https://doi.org/10.3109/17435390.2010.509519

Gecer EN (2021) Green synthesis of silver nanoparticles from *Salvia aethiopis* L. and their antioxidant activity. J Inorgan Organometal Polym Mater 31(11):4402–4409. https://doi.org/10.1007/s10904-021-02057-3

Gour A, Jain NK (2019) Advances in green synthesis of nanoparticles. Artif Cells Nanomed Biotechnol 47(1):844–851. https://doi.org/10.1080/21691401.2019.1577878

Gul A, Fozia Shaheen A, Ahmad I, Khattak B, Ahmad M, Mahmood HM (2021) Green synthesis, characterization, enzyme inhibition, antimicrobial potential, and cytotoxic activity of plant mediated silver nanoparticle using Ricinus communis leaf and root extracts. Biomolecules 11(2):206. https://doi.org/10.3390/biom11020206

Gürmen S, Ebin B, İtü M (2008) Nanopartiküller ve üretim yöntemleri-1. Metalurji Dergisi 150:31–38

Hosseinzadeh K, Roghani S, Mogharrebi AR, Asadi A, Waqas M, Ganji DD (2020) Investigation of cross-fluid flow containing motile gyrotactic microorganisms and nanoparticles over a three-dimensional cylinder. Alex Eng J 59(5):3297–3307. https://doi.org/10.1016/j.aej.2020.04.037

Hulla JE, Sahu SC, Hayes AW (2015) Nanotechnology: history and future. Hum Exp Toxicol 34(12):1318–1321. https://doi.org/10.1177/0960327115603588

Iliger KS, Sofi TA, Bhat NA, Ahanger FA, Sekhar JC, Elhendi AZ, Khan, (2021) Copper nanoparticles: green synthesis and managing fruit rot disease of chilli caused by Colletotrichum capsici. Saudi J Biol Sci 28(2):1477–1486. https://doi.org/10.1016/j.sjbs.2020.12.003

Jeevanandam J, Chan YS, Danquah MK (2016) Biosynthesis of metal and metal oxide nanoparticles. ChemBioEng Rev 3(2):55–67. https://doi.org/10.1002/cben.201500018

Khan I, Saeed K, Khan I (2019) Nanoparticles: properties, applications and toxicities. Arab J Chem 12(7):908–931. https://doi.org/10.1016/j.arabjc.2017.05.011

Köktürk M, Alak G, Atamanalp M (2020) The effects of n-butanol on oxidative stress and apoptosis in zebra fish (Danio rerio) larvae. Comp Biochem Physiol C Toxicol Pharmacol 227:10863

Köktürk M, Çomaklı S, Özkaraca M, Alak G, Atamanalp M (2021) Teratogenic and Neurotoxic Effects of n-Butanol on Zebrafish Development. J Aquat Anim Health 33(2):94–106

Kokturk M, Yıldırım S, Atamanalp M, Calimli MH, Nas MS, Bolat I, Alak G (2022a) Assessment of oxidative DNA damage, apoptosis and histopathological alterations on zebrafish exposed with green silver nanoparticle. Chem Ecol 38(7):655–670. https://doi.org/10.1080/02757540.2022.2108808

Kokturk M, Yıldırım S, Calimli MH, Nas MS, Ibaokurgil F, Ozhan G, Alak G (2023) Perspective on green synthesis of RP-Pd/AC NPs: characterization, embryonic and neuronal toxicity assessment. Int J Environ Sci Technol 20(1):871–882. https://doi.org/10.1007/s13762-022-04005-1

Kokturk M, Yıldırım S, Nas MS, Ozhan G, Atamanalp M, Bolat I, Alak G (2022b) Investigation of the oxidative stress response of a green synthesis nanoparticle (RP-Ag/ACNPs) in zebrafish. Biol Trace Elem Res 200(6):2897–2907. https://doi.org/10.1007/s12011-021-02855-3

Köktürk M, Yildirim S, Yiğit A, Ozhan G, Bolat İ, Alma MH, Atamanalp M (2022) What is the eco-toxicological level and effects of graphene oxide-boramidic acid (GO-ED-BA NP)? In vivo study on Zebrafish embryo/larvae. J Environ Chem Eng 10(5):108443. https://doi.org/10.1016/j.jece.2022.108443

Mehwish HM, Rajoka MSR, Xiong Y, Cai H, Aadil RM, Mahmood Q, Zhu Q (2021) Green synthesis of a silver nanoparticle using *Moringa oleifera* seed and its applications for antimicrobial

and sun-light mediated photocatalytic water detoxification. J Environ Chem Eng 9(4):105290. https://doi.org/10.1016/j.jece.2021.105290
Mutaf T, Bilgin GÇ, Öncel S, Elibol M (2023) Metal nanopartiküllerin mikroalgler aracılığı ile yeşil sentezi. Ege J Fisher Aquat Sci 40(1):81–89. https://doi.org/10.12714/egejfas.40.1.12
Nayem SA, Sultana N, Haque MA, Miah B, Hasan MM, Islam T, Ahammad AS (2020) Green synthesis of gold and silver nanoparticles by using amorphophallus paeoniifolius tuber extract and evaluation of their antibacterial activity. Molecules 25(20):4773. https://doi.org/10.3390/molecules25204773
Primožič M, Knez Ž, Leitgeb M (2021) (Bio) Nanotechnology in food science—food packaging. Nanomaterials 11(2):292. https://doi.org/10.3390/nano11020292
Rana MS, Bhushan S, Prajapati SK (2020) New insights on improved growth and biogas production potential of *Chlorella pyrenoidosa* through intermittent iron oxide nanoparticle supplementation. Sci Rep 10(1):14119. https://doi.org/10.1038/s41598-020-71141-4
Sahoo S, Gopalan A, Ramesh S, Nirmala P, Ramkumar G, Agnes Shifani S, Joshua Ramesh Lalvani JI (2021) Preparation of polymeric nanomaterials using emulsion polymerization. Adv Mater Sci Eng 2021:1–9. https://doi.org/10.1155/2021/1539230
Setia A, Mehata AK, Malik AK, Viswanadh MK, Muthu MS (2023) Theranostic magnetic nanoparticles: synthesis, properties, toxicity, and emerging trends for biomedical applications. J Drug Deliv Sci Technol. https://doi.org/10.1016/j.jddst.2023.104295
Shah SS, Shaikh MN, Khan MY, Alfasane MA, Rahman MM, Aziz MA (2021) Present status and future prospects of jute in nanotechnology: a review. Chem Rec 21(7):1631–1665. https://doi.org/10.1002/tcr.202100135
Tai H, Duan Z, He Z, Li X, Xu J, Liu B, Jiang Y (2019) Enhanced ammonia response of $Ti_3C_2T_x$ nanosheets supported by TiO_2 nanoparticles at room temperature. Sens Actuat B Chem 298:126874. https://doi.org/10.1016/j.snb.2019.126874
Tailor G, Yadav BL, Chaudhary J, Joshi M, Suvalka C (2020) Green synthesis of silver nanoparticles using *Ocimum canum* and their anti-bacterial activity. Biochem Biophys Rep 24:100848. https://doi.org/10.1016/j.bbrep.2020.100848
Thakur A, Sharma N, Bhatti M, Sharma M, Trukhanov AV, Trukhanov SV, Thakur P (2020) Synthesis of barium ferrite nanoparticles using rhizome extract of Acorus calamus: characterization and its efficacy against different plant phytopathogenic fungi. Nano-Struct Nano-Objects 24:100599. https://doi.org/10.1016/j.nanoso.2020.100599
Thakur S, Karak N (2012) Green reduction of graphene oxide by aqueous phytoextracts. Carbon 50(14):5331–5339. https://doi.org/10.1016/j.carbon.2012.07.023
Ucar A, Parlak V, Ozgeris FB, Yeltekin AC, Arslan ME, Alak G, Atamanalp M (2022) Magnetic nanoparticles-induced neurotoxicity and oxidative stress in brain of rainbow trout: mitigation by ulexite through modulation of antioxidant, anti-inflammatory, and antiapoptotic activities. Sci Total Environ 838:155718. https://doi.org/10.1016/j.scitotenv.2022.155718
Ucar A, Yeltekin AÇ, Köktürk M, Calimli MH, Nas MS, Parlak V, Alak G, Atamanalp M (2023) Has PdCu@ GO effect on oxidant/antioxidant balance? Using zebrafish embryos and larvae as a model. Chem Biol Interact 378:110484. https://doi.org/10.1016/j.cbi.2023.110484
Valerio-García RC, Carbajal-Hernández AL, Martínez-Ruíz EB, Jarquín-Díaz VH, Haro-Pérez C, Martínez-Jerónimo F (2017) Exposure to silver nanoparticles produces oxidative stress and affects macromolecular and metabolic biomarkers in the goodeid fish Chapalichthys pardalis. Sci Total Environ 583:308–318. https://doi.org/10.1016/j.scitotenv.2017.01.070
Yavuz İ, Yılmaz EŞ (2021) Biyolojik Sistemli Nanopartiküller. Gazi Üniversitesi Fen Fakültesi Dergisi 2(1):93–108. https://doi.org/10.5281/zenodo.4843592
Zhao Y, Yang X, Yan L, Bai Y, Li S, Sorokin P, Shao L (2021) Biomimetic nanoparticle-engineered superwettable membranes for efficient oil/water separation. J Membr Sci 618:118525. https://doi.org/10.1016/j.memsci.2020.118525
Zhou Y, Wang Y, Peijnenburg W, Vijver MG, Balraadjsing S, Fan W (2023) Using machine learning to predict adverse effects of metallic nanomaterials to various aquatic organisms. Environ Sci Technol. https://doi.org/10.1021/acs.est.2c07039

Chapter 10
Experimental Animal Preference in Aquatic Toxicology

Utku Güner

Abstract Today, the use of experimental animals in scientific research has increased significantly, especially in the last 10 years. The basis of this increase is the increase in the number and species diversity of aquatic organisms as experimental animals for various reasons. Researchers must understand both general experimental animal ethics and principles for aquatic animal experiments and the characteristics of experiments in aquatic environments. This section describes the criteria used in selecting suitable candidates for aquatic experiments. The advantages and disadvantages of different experimental systems are also discussed. However, the questions that should be posed in the selection of aquatic experimental animals and the comparative advantages and disadvantages of different animal species are also presented.

Keywords Aquatic test animals · Fish experiments · Aquatic test systems · Aquatic experimental animal · Toxic experiment design

10.1 Introduction

The use of experimental animals is carried out with different objectives ranging from understanding the biological properties of species to solving the basic mechanisms of diseases, testing cosmetic products and the development of new drugs (Kutluyer and Aksakal 2013). The number of different organisms used in these kinds of researches is constantly increasing today (Erkeç and Arıhan 2014). For instance, in 2005, 11 million vertebrate animal individuals were used in Japan, 12 million in European Union countries and approximately 17 million in the United States of America in researches aiming to analyze physiological mechanisms (Kaya and Çevik 2011).

The researchers of studies in which experimental animals are used are strongly recommended to follow the 3R rule (Reduction, Refinement, Replacement) in their

U. Güner (✉)
Department of Biology, Faculty of Science, Trakya University, Edirne, Türkiye
e-mail: uguner@trakya.edu.tr

M. Atamanalp et al., *Aquatic Toxicology in Freshwater*, Springer Water,
https://doi.org/10.1007/978-3-031-56669-1_10

studies. The number of such rules has recently increased to 7 (Aske and Waugh 2017). At this point, especially in recent years, there has been a tendency to use fewer animals and to use alternative methods (3R principle), which include reducing and improving animal experiments and applying other methods instead of animal use (Festing and Altman 2002). The term "reducing the number of animals used" refers to minimizing the number of animals required for a desired method to be applied, while remaining faithful to the necessity of scientific methods. This approach aims to minimize the use of animals in order to obtain valid results. "Improvement" involves optimizing procedures or experiments performed on animals in ways that cause less pain and distress to the animals. This approach aims to organize the procedures required to obtain scientific data in a way that maximizes animal welfare.

The term "substitution" refers to the use of phylogenetically inferior species or non-animal systems to replace animals traditionally used for experiments or studies. This approach aims to explore alternative methods to achieve scientific objectives and minimize the use of animals. These concepts reflect important principles that aim to maintain scientific progress and protect animal welfare by addressing the ethical and practical aspects of animal use in scientific research (Erkekoglu et al. 2011). Aquatic animals are the most important organisms that meet all these principles to the best extent.

In recent years, the use of aquatic organisms in experimental research has increased. Especially, the use of acute toxicity tests on aquatic organisms in order to evaluate the potential hazards of chemical pollutants has an important place (Kayhan et al. 2014).

Acute toxicity tests performed under static experimental conditions provide rapid and reproducible concentration–response curves to predict the toxic effects of chemicals on aquatic organisms (Güner 2009; Fırat et al. 2011; Guner 2012b). These tests are used to determine the response of species to various chemicals. These experimental studies provide a database for predicting the effects of toxic substances on different organisms. Acute toxicity tests are also frequently used to assess the potential hazards of chemical pollutants (Çoğun et al. 2012; Fırat et al. 2011; Narges et al. 2012). Static acute toxicity experiments provide rapid and reproducible concentration–response curves to estimate the toxic effects of chemicals on aquatic organisms (Rand et al. 1976; Çogun et al. 2012).

This chapter aims to guide researchers by providing basic information about aquatic animal experiments. In particular, it aims to provide guidance to young researchers planning to conduct aquatic animal experiments on experimental design and experimental animal selection. It also aims to provide guidance on how to select appropriate organisms for aquatic animal experiments.

Aquatic animal experimentation refers to the use of organisms living in aquatic environments as experimental material for researches in medical, biological, environmental and other scientific fields. These experiments are carried out to understand the biological responses, behavior, health status and environmental interactions of organisms. Aquatic animals are favored for such experiments because they have several advantages:

1. Access to Different Species: Aquatic environments are habitats for many different species. Therefore, various types of organisms can be found in the same geographical area. Researchers can use this advantage to study similarities and differences between different species.
2. Easy Reproducibility: Many species of aquatic organisms can reproduce rapidly. This makes it easier to provide experimental material on a large scale. Most aquatic organisms can change generations in a short period of time.
3. Controllable Environments: When performing aquatic experiments in laboratory settings, environmental factors such as water temperature, pH levels and water quality can be controlled. This increases the repeatability of the experiments and the reliability of the results.
4. Visual Observation Opportunity: Organisms living underwater are easily visible to observe and record their natural behavior. This provides an ideal environment for behavioral studies.
5. Genetic Diversity: The genetic diversity of different aquatic species provides opportunities for evolutionary and genetic research. Such experiments can contribute to the understanding of genetic variation among organisms.

10.2 Criteria for Selecting Appropriate Organisms for Aquatic Animal Experiments

In aquatic animal experiments, it is necessary to select the appropriate test model and the appropriate experimental organism as well as the experimental design (Rand et al. 1976). Undoubtedly, there are many different factors in the selection of an aquatic animal as an experimental animal (Johnson and Finley 1980; Rand et al. 1976): The most important of these are:

1- Native species representing the ecosystem should be used as much as possible.
2- The species to be selected should be easily available, sufficient in number during the test period and have similar characteristics (age, sex, etc.).
3- Species should have ecological and economic importance.
4- Aquatic animals with a wide range of sensitivity should be selected as much as possible, since sensitivity varies within and between species under appropriate experimental conditions.
5- Aquatic experimental animals should have high adaptability and fast, easy, economical production and culture should be possible.
6- Aquatic experimental animals should be able to live in experimental conditions for 1 week to one month depending on the species.
7- The biology, life requirements, physico-chemical requirements such as salinity, pH and temperature of aquatic experimental animals should be known and provided.
8- It is necessary to know and provide the biology, life requirements, physico-chemical requirements such as salinity, pH and temperature of aquatic experimental animals.

9- The level of the organism in the food chain, its economic importance and the most sensitive stage should be known.
10- Aquatic experimental animals to be used in the experiment should be of appropriate size and weight.
11- In the selection of aquatic experimental animals, experimental containers or aquariums with water at least 100 times the selected animal biomass should be used.
12- In aquatic animals, the basic rules that apply to all experimental animals are applied. The study is planned with the minimum number of animals (to give meaningful results).
13- At the beginning of the study, it is recommended to conduct a pilot study or preliminary experiment before the actual experiment (Festing and Altman 2002).

In all studies using animals, there are basic points that researchers should answer and put forward at the very beginning of the study:

1. Determining which tissues will be used in the research (cell type, tissue, organ, communication between organs, etc.) and how much of these tissues are needed,
2. Determining which animal species / breed they will respond to the research question,
3. Determining which animal species or breed is technically advantageous for these characteristics and which will be least disturbed by the experimental procedures,
4. Determining and reviewing the factors in practice such as procurement, housing, care, adaptation, follow-up, making the infrastructure suitable, literature information, experience with animals,
5. It is necessary to select an appropriate animal model in terms of scientific practice and ethics (Care et al. 1985).

10.3 Aquatic Animal Experimental Systems

Some of the features to be considered in the design of aquatic animal experimental systems may be as follows:

1. **Water Quality and Control:** Water quality is important for experiments to produce reliable results. Factors such as water temperature, pH level and oxygen content should be controlled.
2. **Habitat Simulation:** Simulating conditions similar to the natural habitats of organisms increases the realism of the experimental results.
3. **Appropriate Feed and Nutrition:** Feeding of organisms and feeding regime may affect the experimental results. The type and amount of feed should be well adjusted.
4. **Isolation and Contamination Control:** Experimental systems should minimize the effects of external factors on organisms and prevent external contamination.

5. **Observation and Data Collection:** Appropriate design and equipment are required for observation of organisms and collection of data.
6. **Replicability:** Reproducibility of experimental results ensures scientific validity. Therefore, systems should be designed to be reproducible.

Types of aquatic animal experiments can be classified according to the purpose.

- **Toxicology and Pharmacology Experiments:** Static systems are ideal for short-term toxicity tests.
- **Ecological and Behavioral Studies:** Fluid or open systems can be used to study the natural behavior of organisms and ecosystem interactions.
- **Genetic and Evolutionary Studies:** Closed or open systems can be used to understand genetic variation and evolutionary processes. In general, which experimental system to choose should be determined by the objectives of the experiment, the characteristics of the organisms and the needs of the researcher. Depending on the research topic and the animal species being studied, a variety of aquatic experimental systems are available. In general, there are four basic experimental systems used in aquatic animal experiments.
- **Static experiment system:** It is the experimental system in which the water and the applied chemical are not changed during the experiment (see Fig. 10.1).
- **Recirculation experimental system:** A pump is used to circulate the water and chemicals used in the experiment in a continuous loop. This circulation system allows the water and chemicals to pass through a filter or similar system before returning to the experimental aquarium. This mechanism is particularly important in experiments that aim to elucidate the effect of the chemical under investigation. However, in this system, the problem may arise that the chemical under investigation remains in the filter, accumulates and does not return to the experimental aquarium due to the effect of the filter (such as an aerator or sterilizer) (see Fig. 10.1).
- **Regenerative experimental system:** In this aquatic experimental system, a static experimental system is used and the experimental conditions are changed with fresh water at regular intervals (usually every 24 h). This approach ensures that the aquatic animal is continuously exposed to the same amount of chemicals, while at the same time removing the metabolites released by the organism into the environment (see Fig. 10.1).

The selection of aquatic animal experimental systems is a process that should be compatible with the research subject, serve the purpose of the study, be easily applicable, be economical, simple and controllable. Each aquatic experimental system has its own advantages and disadvantages. The comprehensive advantages and disadvantages of these systems are summarized in Table 10.1.

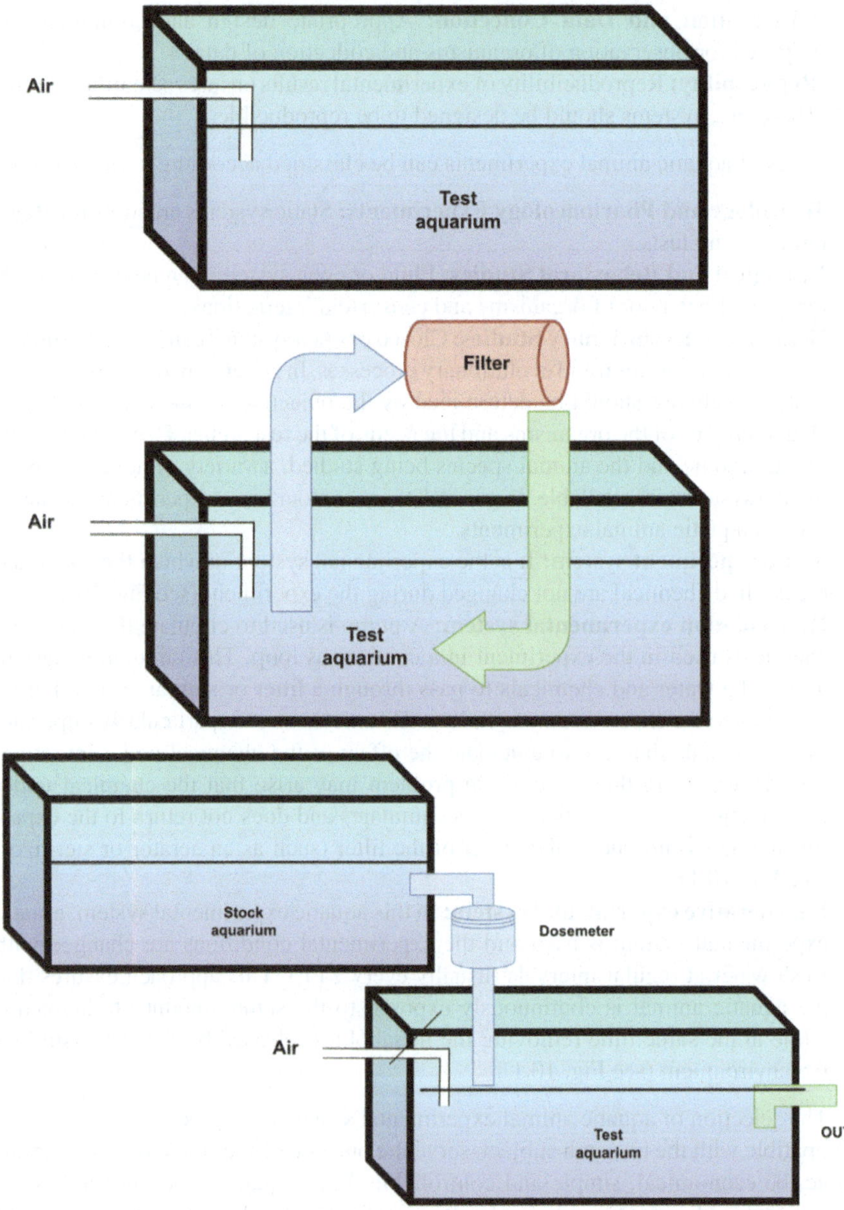

Fig. 10.1 Basic design of the Static test system, Recirculation test system and flow test system from top to bottom

Table 10.1 Advantages and disadvantages of aquatic experimental systems (Güner 2018)

Experimental system	Cost Substance	Water consumption	Experience	Metabolite accumulation	Stress	Stability of environmental condition	Long-term operation	Special infrastructure
Static	+	+	+	+++	+	+	+	+
Recirculation	+++	+	+++	+	+	+++	+++	+++
Renewal	+	++	+	++	+++	++	++	+
Flow	+++	+++	+++	+	+	+++	+++	+++

+: very difficult and/or very little, ++: feasible and/or in sufficient quantity, ++++:easy and/or in large quantity

10.3.1 The Most Widely Used Aquatic Experimental Animals

Many different aquatic organisms are used for experimental purposes. Undoubtedly, different organisms can be selected according to the question to be answered by the research. The characteristics of the organism to be selected are the most important points in the research. Freshwater experimental animals are the preferred organisms for various scientific researches. Some of the most preferred freshwater experimental animals:

1. **Daphnia** (Water Flea): Used for water quality tests, toxicity experiments and ecosystem health assessment. Rapid reproduction and visual observation are among their advantages.
2. **Zebrafish** (*Danio rerio*): Widely used in fields such as genetics, neurology, developmental biology and drug testing. Spawning and embryo development is fast.
3. **Medaka** (*Oryzias latipes*): Preferred for genetic and developmental studies like zebrafish. There are different genotypes and color variations.
4. **Crayfish** (*Astacus* spp.): Used for ecosystem health assessment, water quality analyses and toxicity tests. They play an important role in ecological experiments.
5. **Ampionts** (*Hydra* spp.): They are favored for their regeneration ability and understanding of cellular mechanisms.
6. **Crabs** (*Procambarus* spp.): Used for behavioral studies, biochemical analyses and water quality experiments.
7. **Mussels** (*Dreissena* spp.): Used for water pollution monitoring, filtration and studies on ecosystem interactions.
8. **Fish** (e.g. Guppy, *Poecilia reticulata*): Preferred for the study of genetic variation, behavioral analyses and toxicity tests.

These organisms are selected according to the aims of the experiments and the needs of the researchers. Each type of organism has specific biological characteristics, advantages and limitations.

Experimental animal selection is a process in which various factors are analyzed and these factors vary in different studies. Parameters such as sensitivity, adaptability, native species status, cost, number of individuals required, potential to obtain tissue or blood/haemolymph are the main determinants of organism selection in aquatic animal experiments. How these parameters are evaluated in different study contexts is summarized in Fig. 10.2.

The design phase of aquatic animal experiments includes the planning and preparation phase prior to the experiment. The main questions that the researcher needs to address at this stage are presented in below for the purpose of the experiment and organism selection.

The aim of this experimental design process is to ensure that the appropriate organism is selected and to increase the reliability of the experiment and the validity of the results. This stage contributes to the effective and scientifically based realization of aquatic animal experiments.

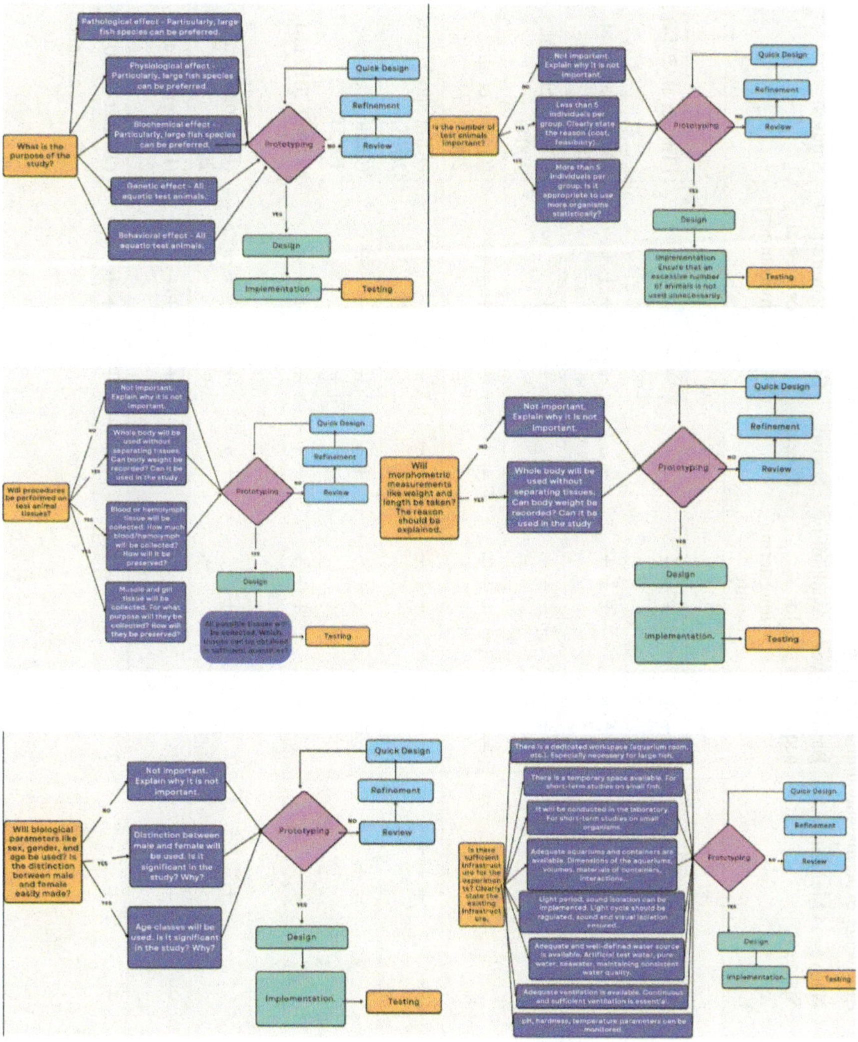

Fig. 10.2 Basic algorithms in aquatic animal experiment design

Experimental animal selection is a process in which various factors are analyzed and these factors vary in different studies. Parameters such as sensitivity, adaptability, native species status, cost, number of individuals required, potential to obtain tissue or blood/haemolymph are the main determinants of organism selection in aquatic animal experiments. How these parameters are evaluated in different study contexts is summarized in Table 10.2.

The design phase of aquatic animal experiments includes the planning and preparation phase prior to the experiment. The main questions that the researcher needs

Table 10.2 Advantages and disadvantages of freshwater experimental animals compared to each other (Güner 2018)

	Sensitivity	Adaptation	Native Species	Maintenance	Blood-tissue acquisition	Availability for long-term study	Native Species	Cost	Requirement for special environment	Studies
Vertebrate species										
Salmo sp.	+++	+	+	+	+++	+++	+	+++	+++	(Türe et al. 2014)
Oncorhynchus mykiss	+++	+	+	+	+++	+++	+	+++	+++	(Atamanalp et al. 2003)
Cyprinus carpio	+	+++	+	+++	+++	+++	+	+	+	(De Boeck et al. 2004)
Danio rerio	++	+++	–	+++	++	+++	+++	+++	+++	(Griffitt et al. 2007)
Oreochromis niloticus	+	++	–	+++	+++	+++	++	++	+	(Alim and Matter 2015)
Poecilia sp.	++	++	-	++	++	+++	+++	+++	+++	(Gallo et al. 1995)
Gambusia sp.	+	+++	–	++	++	+++	+++	+	+++	(Guner 2012a,b)
Invertebrate species										
Artemia sp.	+++	+	–	+	?	+	+++	+++	+++	(Dağlıoğlu et al. 2016)
Daphnia sp	+++	+	+	+	?	+	+++	+	++	(Lavorgna et al. 2016)
Gammarus sp.	+++	++	+	+	?	+	+++	+	++	(Türe et al. 2014)

(continued)

Table 10.2 (continued)

	Sensitivity	Adaptation	Native Species	Maintenance	Blood-tissue acquisition	Availability for long-term study	Native Species	Cost	Requirement for special environment	Studies
Ostracoda	+++	+	+	+	?	+	+++	+	+++	(Sevilla et al. 2013)
Astacus sp.	++	+	+	+	+++	++	+	++	+++	(Güner 2010)

+: very difficult and/or very little, ++: feasible and/or in sufficient quantity, ++++: easy and/or in sufficient quantity, ?: impossible to address at this stage are presented in Table 10.2 for the purpose of the experiment and organism selection.

The aim of this experimental design process is to ensure that the appropriate organism is selected and to increase the reliability of the experiment and the validity of the results. This stage contributes to the effective and scientifically based realization of aquatic animal experiments.

The aim of answering these questions is to improve the experimental design and enable the researcher to conduct experiments using the correct method and appropriate organisms (Fig. 10.2). The questions to be addressed in the experimental design phase of the research are crucial for obtaining accurate and precise data at the end of the experiment.

10.4 Conclusion

It is known that the use of large numbers of organisms in aquatic animal experiments may lead to ethical problems. However, the use of a sufficient number of organisms is critical for the statistical accuracy and sensitivity of the study The correct number of experimental animals is determined according to the 3R principle (Reduction, Refinement, Replacement and also 7R rules), aiming to obtain statistically correct results with minimal animal use (Festing and Altman 2002).

The organism to be selected in aquatic animal experiments has an important role in determining the design of the study and the system (Fig. 10.1, Static, Recirculation and Flow experimental systems). The choice of organism is determined by factors such as the location, purpose and by whom the research is conducted. The characteristics and adaptability of the organism affect the suitability of the experiment for the purpose. Different aquatic animal species may exhibit advantages and disadvantages under different experimental conditions.

For example, LC_{50} studies using small organisms (e.g. *Gammarus*, *Daphnia*, *Artemia*) have the advantages of small volumes, more repetitions and larger numbers of individuals, whereas studies with large organisms (e.g. carp) may require a larger laboratory infrastructure. For long-term studies requiring tissue, blood or haemolymph collection, the use of larger fish or invertebrate organisms may provide the advantage of collecting more tissue samples.

The choice of organism should be based on factors such as the location of the experiment, the purpose of the experiment, the level of experience of the researcher and available resources. Organisms that are available from natural sources, can be cultivated and are suitable for the needs should be preferred. The life cycle, behavior and biological characteristics of these organisms should be well understood.

Consequently, it is of great importance to carefully select the organism at the beginning of the study, to assess its suitability for the objectives of the experiment, its advantages and disadvantages. Organism selection is a fundamental step in the design of the study and the reliability of the results. Conducting fast, effective and ethical aquatic animal experiments will fulfil the requirements of the 3R (or 7R) principle.

Acknowledgements Many thanks to my professors who have been the basis for me in aquatic toxicology and aquatic animal experiments, my colleagues in the Department of Biology who supported me in my research, my graduate students for their help, support and interest in my research, and Trakya University for their economic support in my academic studies.

References

Alim D, Matter H (2015) Histopathological alteration induced in gills of juvenile Nile Tilapia *Oreochromis niloticus* upon exposure to two bio-pesticides. Int J Fisheries and Aquatic Stud 2(5):80–83

Anonymous (2014) Regulation on working procedures and principles of animal experiments ethics committees. Official Gazette Issue: 28914. http://www.resmigazete.gov.tr/eskiler/2014/02/20140215-6.htm

Aske KC, Waugh CA (2017) Expanding the 3R principles: more rigour and transparency in research using animals. EMBO Reports 18.9:1490–1492. https://doi.org/10.15252/embr.201744428

Atamanalp M, Bayır A, Sirkecioğlu AN, Cengiz M (2003) The effects of sublethal doses a disinfectant (Malachite Green) on blood parameters of rainbow trout (*Oncorhynchus mykiss*). J Gazi Univer Gazi Educat Faculty 23(3):177–187

De Boeck G, Meeus W, De Coen W, Blust R (2004) Tissue-specific Cu bioaccumulation patterns and differences in sensitivity to waterborne Cu in three freshwater fish: rainbow trout (*Oncorhynchus mykiss*), common carp (*Cyprinus carpio*), and gibel carp (*Carassius auratus gibelio*). Aquatic Toxicol 70(3):179–188. http://www.sciencedirect.com/science/article/pii/S0166445X04002231

Care IoLARCo, Animals UoL, Resources NIoHDoR (1985) Guide for the care and use of laboratory animals. National Academies

Çogun HY, Fırat Ö, Fırat Ö, Yüzereroğlu TA, Gök G, Kargin F, Kötemen Y (2012) Protective effect of selenium against mercury-induced toxicity on hematological and biochemical parameters of *Oreochromis niloticus*. J Biochem Mol Toxicol 26(3):117–122. https://doi.org/10.1002/jbt.20417

Çoğun HY, Çapar SÖ, Çağlar R, Taşyürek K, Tanrıver B, Özdemir S, Er CB, Çimrin İ, Sarıçicek E (2013) Toxic effects of dimethoat on some enzyme systems in *Oreochromis niloticus* (Linnaeus, 1758) Anatolian J Natural Sci 4.2:33–36

Dağlıoğlu Y, Çelebi MS, Önalan Ş (2016) Determination of acute toxic effects of poly (Vinylferrocenium) supported palladium nanoparticle (Pd/PVF+) on *Artemia salina*. Pakistan J Zool 48(1):187–193

Erkeç ÖE, Arıhan O (2014) Animal models in medical studies. Int J Hum Sci 11(2):50–63

Erkekoglu P, Giray BK, Başaran N (2011) 3R principle and alternative toxicity testing methods. J Pharm Sci 36:105–118

Festing MF, Altman DG (2002) Guidelines for the design and statistical analysis of experiments using laboratory animals. ILAR J 43(4):244–258. https://doi.org/10.1093/ilar.43.4.244

Fırat Ö, Cogun HY, Yüzereroğlu TA, Gök G, Fırat Ö, Kargin F, Kötemen Y (2011) A comparative study on the effects of a pesticide (cypermethrin) and two metals (copper, lead) to serum biochemistry of Nile tilapia *Oreochromis Niloticus*. Fish Physiol Biochem 37(3):657–666. https://doi.org/10.1007/s10695-011-9466-3

Gallo D, Merendino A, Keizer J, Vittozzi L (1995) Acute toxicity of two carbamates to the guppy (*Poecilia reticulata*) and the zebrafish (*Brachydanio rerio*). Sci Total Environ 171(1):131–136. https://doi.org/10.1016/0048-9697(95)04681-X

Griffitt RJ, Weil R, Hyndman KA, Denslow ND, Powers K, Taylor D, Barber DS (2007) Exposure to copper nanoparticles causes gill injury and acute lethality in zebrafish (*Danio rerio*). Environ Sci Technol 41(23):8178–8186. http://pubs.acs.org/doi/pdfplus/https://doi.org/10.1021/es071235e

Güner U (2009) Determination of lambda-cyhalotrin (Tekvando 5EC) 96 hour lethal concentration 50 at *Gambusia affinis* (Baird and Girard, 1853). J Fish Sci 3(3):214–219

Güner U (2010) Heavy metal effects on P, Ca, Mg, and total protein contents in embryonic pleopodal eggs and stage-1 juveniles of freshwater crayfish *Astacus leptodactylus* (Eschscholtz, 1823). Turk J Biol 34(4):405–412. https://doi.org/10.3906/biy-0811-19

Güner U (2018) Selection of aquatic experimental animals, advantages and disadvantages. J Instit Sci Technol 8(3):93–100. https://doi.org/10.1073/pnas.0611690104

Guner U (2012a) Alüminyumun, Farklı pH Değerlerinde *Gambusia affinis* (Baird and Girard, 1853) Üzerine Toksisitesi. In: Conference: V. national limnology symposium. https://doi.org/10.13140/2.1.1709.6004

Guner U (2012b) Determination of 24-and 48-hour LC_{50} values of diazinon in *Gambusia affinis* (baird and girard, 1853). Archives of Indus Hygiene and Toxicol 63(2):32–33. https://doi.org/10.22392/egirdir.292971

Johnson WW, Finley MT (1980) Handbook of acute toxicity of chemicals to fish and aquatic invertebrates: summaries of toxicity tests conducted at Columbia National Fisheries Research Laboratory, 1965–78. US Fish and Wildlife Service

Kaya M, Çevik A (2011) Planning in animal experiments and choosing model. Experimed 2:36–39

Kayhan F, Kaymak G, Akbulut C, Esmer H, Yön ND (2014) Oxidative stress mechanisms and adaptive responses to environmental conditions in aquatic organisms. Marmara J Sci 26(4):137–151

Kutluyer F, Aksakal E (2013) Sucul model organizmalar ve biyoteknolojide kullanimi. Anadolu Tarım Bilimleri Dergisi 28(2):101–107

Lavorgna M, Russo C, D'Abrosca B, Parrella A, Isidori M (2016) Toxicity and genotoxicity of the quaternary ammonium compound benzalkonium chloride (BAC) using *Daphnia magna* and *Ceriodaphnia dubia* as model systems. Environ Pollut 210:34–39. https://doi.org/10.1016/j.envpol.2015.11.042

Narges AB, Ahmad S, Mohammadsedigh M, Hossein Z, Negin S, Zohreh AB (2012) Non-essential metals (Cd and Pb) accumulation and elimination in liver tissue of juvenile milkfish, after sublethal exposure. Indian J Geo-Marine Sci 41(5):412–417

Rand M, Greenberg AE, Taras MJ (1976) Standard methods for the examination of water and wastewater. Prepared and published jointly by American Public Health Association, American Water Works Association, and Water Pollution Control Federation

Rand GM (1995) Fundamentals of aquatic toxicology: effects, environmental fate and risk assessment. CRC Press

Sevilla JB, Nakajima F, Yamamoto K (2013) Effect of food and light on the sensitivity of copper and zinc to freshwater benthic ostracod *Heterocypris incongruens*. J Water and Environ Technol 11(3):249–261. https://www.jstage.jst.go.jp/article/jwet/11/3/11_249/_pdf

Türe M, Haliloğlu Hİ, Altuntaş C, Boran H, Kutlu İ (2014) Comparison of experimental susceptibility of rainbow trout (Oncorhynchus mykiss), Turbot (Psetta maxima), Black Sea Trout (Salmo trutta labrax) and Sea Bass (Dicentrarchus labrax) to Lactococcus garvieae. Turk J Fish Aquat Sci 14:507–513

SPRINGER NATURE

GPSR Compliance

The European Union's (EU) General Product Safety Regulation (GPSR) is a set of rules that requires consumer products to be safe and our obligations to ensure this.

If you have any concerns about our products, you can contact us on ProductSafety@springernature.com

In case Publisher is established outside the EU, the EU authorized representative is:

Springer Nature Customer Service Center GmbH
Europaplatz 3
69115 Heidelberg, Germany

The manufacturer's authorised representative in the EU is Springer Nature Customer Service Centre GmbH, Europaplatz 3, 69115 Heidelberg, Germany. If you have any concerns regarding our products, please contact ProductSafety@springernature.com

Printed and bound by CPI Group (UK) Ltd, Croydon, CR0 4YY

28/09/2026

02224081-0001